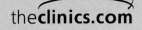

RADIOLOGIC

CLINICS

OF NORTH AMERICA

Postoperative Musculoskeletal Imaging

Guest Editor

LAURA W. BANCROFT, MD

May 2006 • Volume 44 • Number 3

ELSEVIER
SAUNDERS

An imprint of Elsevier, Inc
PHILADELPHIA LONDON TORONTO MONTREAL SYDNEY TOKYO

W.B. SAUNDERS COMPANY
A Division of Elsevier Inc.

1600 John F. Kennedy Boulevard • Suite 1800 • Philadelphia, Pennsylvania 19103-2899

http://www.theclinics.com

RADIOLOGIC CLINICS OF NORTH AMERICA Volume 44, Number 3
May 2006 ISSN 0033-8389, ISBN 1-4160-3544-3

Editor: Barton Dudlick

Reprints: For copies of 100 or more, of articles in this publication, please contact the Commercial Reprints Department, Elsevier Inc., 360 Park Avenue South, New York, New York 10010-1710. Tel.: (+1) 212-633-3813; Fax: (+1) 212-462-1935; E-mail: reprints@elsevier.com.

The ideas and opinions expressed in *Radiologic Clinics of North America* do not necessarily reflect those of the Publisher. The Publisher does not assume any responsibility for any injury and/or damage to persons or property arising out of or related to any use of the material contained in this periodical. The reader is advised to check the appropriate medical literature and the product information currently provided by the manufacturer of each drug to be administered to verify the dosage, the method and duration of administration, or contraindications. It is the responsibility of the treating physician or other health care professional, relying on independent experience and knowledge of the patient, to determine drug dosages and the best treatment for the patient. Mention of any product in this issue should not be construed as endorsement by the contributors, editors, or the Publisher of the product or manufacturers' claims.

Radiologic Clinics of North America (ISSN 0033-8389) is published in January, March, May, July, September, and November by W.B. Saunders, 360 Park Avenue South, New York, NY 10010-1710. Business and editorial offices: 1600 John F. Kennedy Boulevard, Suite 1800, Philadelphia, Pennsylvania 19103-2899. Accounting and circulation offices: 6277 Sea Harbor Drive, Orlando, FL 32887-4800. Periodicals postage paid at New York, NY, and additional mailing offices. Subscription prices are USD 235 per year for US individuals, USD 350 per year for US institutions, USD 115 per year for US students and residents, USD 275 per year for Canadian individuals, USD 430 per year for Canadian institutions, USD 320 per year for international individuals, USD 430 per year for international institutions and USD 155 per year for Canadian and foreign students/residents. To receive student and resident rate, orders must be accompanied by name of affiliated institution, date of term, and the signature of program/residency coordinator on institution letterhead. Orders will be billed at individual rate until proof of status is received. Foreign air speed delivery is included in all Clinics subscription prices. All prices are subject to change without notice. **POSTMASTER:** Send address changes to *Radiologic Clinics of North America*, Elsevier Periodicals Customer Service, 6277 Sea Harbor Drive, Orlando, FL 32887-4800. **Customer Service: 1-800-654-2452 (US). From outside of the US, call (+1) 407-345-4000.**

Radiologic Clinics of North America also is published in Greek by Paschalidis Medical Publications, Athens, Greece.

Radiologic Clinics of North America is covered in *Index Medicus, EMBASE/Excerpta Medica, Current Contents/Life Sciences, Current Contents/Clinical Medicine, RSNA Index to Imaging Literature, BIOSIS, Science Citation Index,* and *ISI/BIOMED.*

Printed in the United States of America.

GOAL STATEMENT

The goal of the *Radiologic Clinics of North America* is to keep practicing radiologists and radiology residents up to date with current clinical practice in radiology by providing timely articles reviewing the state of the art in patient care.

ACCREDITATION

The *Radiologic Clinics of North America* is planned and implemented in accordance with the Essential Areas and Policies of the Accreditation Council for Continuing Medical Education (ACCME) through the joint sponsorship of the University of Virginia School of Medicine and Elsevier. The University of Virginia School of Medicine is accredited by the ACCME to provide continuing medical education for physicians.

The University of Virginia School of Medicine designates this educational activity for a maximum of 15 AMA PRA Category 1 Credits™. Physicians should only claim credit commensurate with the extent of their participation in the activity.

The American Medical Association has determined that physicians not licensed in the US who participate in this CME activity are eligible for 15 AMA PRA Category 1 Credits™.

Category 1 credit can be earned by reading the text material, taking the CME examination online at http://www.theclinics.com/home/cme, and completing the evaluation. After taking the test, you will be required to review any and all incorrect answers. Following completion of the test and evaluation, your credit will be awarded and you may print your certificate.

FACULTY DISCLOSURE/CONFLICT OF INTEREST

The University of Virginia School of Medicine, as an ACCME accredited provider, endorses and strives to comply with the Accreditation Council for Continuing Medical Education (ACCME) Standards of Commercial Support, Commonwealth of Virginia statutes, University of Virginia policies and procedures, and associated federal and private regulations and guidelines on the need for disclosure and monitoring of proprietary and financial interests that may affect the scientific integrity and balance of content delivered in continuing medical education activities under our auspices.

The University of Virginia School of Medicine requires that all CME activities accredited through this institution be developed independently and be scientifically rigorous, balanced and objective in the presentation/discussion of its content, theories and practices.

All authors/editors participating in an accredited CME activity are expected to disclose to the readers relevant financial relationships with commercial entities occurring within the past 12 months (such as grants or research support, employee, consultant, stock holder, member of speakers bureau, etc.). The University of Virginia School of Medicine will employ appropriate mechanisms to resolve potential conflicts of interest to maintain the standards of fair and balanced education to the reader. Questions about specific strategies can be directed to the Office of Continuing Medical Education, University of Virginia School of Medicine, Charlottesville, Virginia.

The authors/editors listed below have identified no financial or professional relationships for themselves or their spouse/partner:
Mark C. Adkins, MD; Suzanne Anderson, MD; Laura W. Bancroft, MD; Douglas P. Beall, MD; Francesca D. Beaman, MD; Diane Bergin, MD; Thomas H. Berquist, MD; Scot E. Campbell, MD, MAJ, MC, USAF; Mark S. Collins, MD; Barton Dudlick, Acquisitions Editor; Matthew A. Frick, MD; Mark J. Kransdorf, MD; Justin Q. Ly, MD; Hal D. Martin, DO; William B. Morrison, MD; Mark D. Murphey, MD; Jeffrey J. Peterson, MD; Kimberly A. Ruzek, MD; Carolyn M. Sofka, MD; and, Moritz Tannast, MD.

Disclosure of Discussion of Non-FDA Approved Uses for Pharmaceutical and/or Medical Devices. **The University of Virginia School of Medicine, as an ACCME provider, requires that all authors identify and disclose any "off label" uses for pharmaceutical and medical device products. The University of Virginia School of Medicine recommends that each physician fully review all the available data on new products or procedures prior to clinical use.**

TO ENROLL

To enroll in the Radiologic Clinics of North America Continuing Medical Education program, call customer service at 1-800-654-2452 or sign up online at http://www.theclinics.com/home/cme. The CME program is available to subscribers for an additional annual fee USD 205.

POSTOPERATIVE MUSCULOSKELETAL IMAGING

GUEST EDITOR

LAURA W. BANCROFT, MD
Associate Professor of Radiology, Mayo Clinic
Jacksonville, Florida

CONTRIBUTORS

MARK C. ADKINS, MD
Assistant Professor of Radiology, Department
of Radiology; Chair, Division of Musculoskeletal
Radiology, Mayo Clinic and Foundation,
Rochester, Minnesota

SUZANNE ANDERSON, MD
Musculoskeletal Radiology, Department
of Diagnostic, Interventional and Pediatric,
University Hospital of Bern Switzerland,
Bern, Switzerland

LAURA W. BANCROFT, MD
Associate Professor of Radiology, Mayo Clinic
Jacksonville, Florida

DOUGLAS P. BEALL, MD
Department of Radiology, Oklahoma Sports
Science and Orthopaedics; Associate Professor
of Orthopedics, University of Oklahoma,
Oklahoma City, Oklahoma

FRANCESCA D. BEAMAN, MD
Department of Radiology, Mayo Clinic,
Jacksonville, Florida

DIANE BERGIN, MD
Assistant Professor of Radiology, Department of
Radiology, Thomas Jefferson University Hospital,
Philadelphia, Pennsylvania

THOMAS H. BERQUIST, MD, FACR
Professor of Diagnostic Radiology, Mayo Clinic
College of Medicine, Rochester, Minnesota;
Consultant in Diagnostic Radiology,
Department of Radiology, Mayo Clinic,
Jacksonville, Florida

SCOT E. CAMPBELL, MD
Department of Radiology and Nuclear Medicine,
Wilford Hall Medical Center, Lackland AFB, Texas

MARK S. COLLINS, MD
Assistant Professor of Radiology, Department of
Radiology, Division of Musculoskeletal Radiology,
Mayo Clinic and Foundation, Rochester, Minnesota

MATTHEW A. FRICK, MD
Instructor, Department of Radiology, Division of
Musculoskeletal Radiology, Mayo Clinic and
Foundation, Rochester, Minnesota

MARK J. KRANSDORF, MD
Professor of Radiology, Mayo Clinic College of
Medicine, Rochester, Minnesota; Consultant,
Department of Radiology, Mayo Clinic,
Jacksonville, Florida; Visiting Professor,
Department of Radiologic Pathology, Armed
Forces Institute of Pathology, Washington, DC

JUSTIN Q. LY, MD
Department of Radiology and Nuclear Medicine,
Wilford Hall Medical Center, Lackland AFB, Texas

HAL D. MARTIN, DO
Oklahoma Sports Science and Orthopaedics,
Oklahoma City, Oklahoma

WILLIAM B. MORRISON, MD
Associate Professor of Radiology, Chief of
Musculoskeletal Radiology, Thomas Jefferson
University Hospital, Philadelphia, Pennsylvania

MARK D. MURPHEY, MD
Chief, Musculoskeletal Radiology, Department of
Radiologic Pathology, Armed Forces Institute of
Pathology, Washington, DC; Professor of Radiology,
Department of Radiology and Nuclear Medicine,
Uniformed Services University of the Health
Sciences, Bethesda, Maryland; Department of
Radiology, Walter Reed Army Medical Center,
Washington, DC

JEFFREY J. PETERSON, MD
Assistant Professor, Department of Radiology,
Mayo Clinic, Jacksonville, Florida

KIMBERLY A. RUZEK, MD
Musculoskeletal Fellow, Radiology, Mayo Clinic
Jacksonville, Florida

CAROLYN M. SOFKA, MD
Associate Professor of Radiology, Weill Medical
College of Cornell University; Associate Attending
Radiologist, Department of Radiology and Imaging,
Hospital for Special Surgery, New York, New York

MORITZ TANNAST, MD
Department of Orthopaedic Surgery, University of
Bern, Inselspital, Murtenstrasse, Bern, Switzerland

POSTOPERATIVE MUSCULOSKELETAL IMAGING

Volume 44 • Number 3 • May 2006

Contents

of the surgical procedures involve the placement of metallic joint replacements or fixation that can make the imaging evaluation of the postoperative anatomy challenging. Clinical examination of patients combined with the type of procedure performed direct the appropriate imaging evaluation; adequate clinical knowledge of these procedures and how to optimally image them provide an opportunity to attain the most accurate evaluation possible.

The knee is a frequently injured joint and, thus, a common focus of operative intervention. As operative techniques and imaging modalities evolve, radiologists must be aware of the expected postoperative appearance after knee surgeries that are performed commonly and also must be comfortable recognizing complications encountered commonly in the immediate and delayed postoperative period. Drawing on the large amount of attention this subject has received of late in the radiologic and orthopedic literature, this article reviews the knee surgeries performed most commonly and the expected normal and most frequently encountered abnormal postoperative imaging findings with an emphasis on MR imaging.

This article describes the relevant surgical detail and MR imaging appearance of common operations performed in the foot and ankle. To evaluate postsurgical patients critically, it is important to understand the primary clinical diagnosis, surgical treatment undergone, the interval since surgery, and patients' current clinical symptoms. Radiography is the most common imaging modality for evaluation of the postoperative ankle and foot. MR imaging may be useful for evaluating the soft tissues and osseous structures in the postsurgical foot and ankle.

Spinal instrumentation techniques have expanded dramatically during the past several decades, but the search for the perfect operative approach and fixation system continues. Fixation devices are designed for the cervical, thoracic, lumbar, and sacral segments using anterior, posterior, transverse, videoarthroscopic, and combined approaches. In most cases, bone grafting also is performed, because instrument failure occurs if solid bony fusion is not achieved. Radiologists must understand the operative and instrumentation options. Knowledge of expected results, appearance of graft material, and different forms of instrumentation is critical for evaluating position of implants and potential complications associated with operative approaches and spinal fixation devices.

Joint replacement procedures have improved dramatically during the past 30 years fueled by the changes in techniques for hips and knees. Joint replacements in other anatomic regions also have become more popular. It is essential to understand the importance of pre- and postoperative imaging for evaluating patients. Preoperative images are used in concert with clinical data to select the appropriate patients and components. Postoperative imaging is critical for evaluating position and potential complications. Appropriate selection of imaging modalities is essential to provide optimal, cost-effective patient care.

Infectious disease complicating surgery involving the musculoskeletal system is one of the most important causes of postoperative morbidity and mortality. Timely detection and accurate localization of infectious processes have important clinical implications and are critical to appropriate patient management. Imaging studies can play an important role in the detection of infection and can help guide appropriate clinical management. The diagnosis of postoperative infection can be made by a variety of imaging modalities. This article reviews the various methods and modalities available for the detection of postoperative infection.

Bone graft materials quickly are becoming a vital tool in reconstructive orthopedic surgery and demonstrate considerable variability in their imaging appearance. Functions of bone graft materials include promoting osseous ingrowth and bone healing, providing a structural substrate for these processes, and serving as a vehicle for direct antibiotic delivery. The three primary types of bone graft materials are allografts, autografts, and synthetic bone graft substitutes.

As the radiologic evaluation of soft tissue masses has changed dramatically with the advent of MR imaging, the effect of MR imaging is even more striking in the assessment of patients after treatment. In cases of local tumor recurrence, MR imaging has become the standard of care. Using a few basic principles, even small local recurrences can be detected accurately, and recurrence can be distinguished from postoperative or post-treatment change. This review presents a fundamental approach to the evaluation of patients, following treatment for soft tissue tumors and highlighting MR imaging.

RADIOLOGIC
CLINICS
OF NORTH AMERICA

Radiol Clin N Am 44 (2006) xi

Preface
Postoperative Musculoskeletal Imaging

Laura W. Bancroft, MD
Guest Editor

Laura W. Bancroft, MD
Mayo Clinic Jacksonville
4500 San Pablo Road
Jacksonville, FL 32224, USA

E-mail address:
Bancroft.Laura@Mayo.edu

This issue of the *Radiologic Clinics of North America* serves as a centralized resource for radiologists who evaluate postoperative musculoskeletal images. There are numerous musculoskeletal surgical procedures, fixation, instrumentation, and joint replacement systems that can, at times, be overwhelming for the interpreting radiologist. My colleagues and I frequently explore the radiologic and orthopedic journals, books, and the Internet in an attempt to stay current with these devices. When a specific radiologic question arises, we often thumb through such specialty books as Tom Berquist's *Imaging Atlas of Orthopedic Appliances and Prostheses*. It seems, however, that few focused references address multiple facets of postoperative musculoskeletal imaging.

In this issue of *Radiologic Clinics* multiple experienced musculoskeletal radiologists offer insight into optimizing imaging techniques and correctly interpreting postoperative imaging of specific anatomic regions. In addition, the authors address topics that are sparse in the radiologic literature, such as imaging after tumor resection and imaging of bone graft materials.

I would like to extend my gratitude to all of the authors who have worked hard to contribute to this issue. Their willingness to share their expertise is appreciated. In addition, I would like to thank Barton Dudlick and the staff at Elsevier who have made this issue of *Radiologic Clinics of North America* possible.

doi:10.1016/j.rcl.2006.03.001

ELSEVIER
SAUNDERS

RADIOLOGIC
CLINICS
OF NORTH AMERICA

Radiol Clin N Am 44 (2006) 323–329

Optimizing Techniques for Musculoskeletal Imaging of the Postoperative Patient

Carolyn M. Sofka, MD[a,b,*]

- Plain film radiographs
- Nuclear medicine scintigraphy
- Ultrasound
- CT

- MR imaging
- Summary
- References

Surgical techniques and approaches to musculoskeletal conditions have changed dramatically during the past few years; paralleling this has been the need to image these patients after surgery accurately and reliably. Patients can present for imaging after surgical intervention with a variety of symptoms, including vague, nonlocalizing, complaints of pain and stiffness, or they may have experienced a trauma subsequent to the surgery. A general understanding of the surgical procedures performed and knowledge of the appropriate applications and modifications of imaging techniques are needed to diagnose pathology accurately and confidently in postoperative patients.

This review discusses the parameter modifications needed to image postoperative patients confidently with CT and MR imaging, with an emphasis on imaging patients who have painful arthroplasty. The usefulness of plain film radiographs, ultrasound, and nuclear medicine also are discussed.

Plain film radiographs

Plain film radiographs remain the mainstay of diagnostic imaging of the musculoskeletal system. Despite nonlocalizing signs or symptoms, patients may have an abnormality post surgery, such as a fracture, that may be identified clearly on conventional radiographs, obviating further imaging. Patients in a postoperative setting often are non–weight bearing on the affected joint or limb, and the affected joint or limb may become regionally osteopenic. The initial application of stress to an affected area postoperatively can lead to an insufficiency, or stress fracture. At minimum, two views of the affected area, ideally at 90° to one another, should be obtained.

Causes of pain in painful arthroplasty can include infection, osteolysis or aseptic loosening, or fracture. When imaging patients who have implantable orthopedic devices, the entire device or prosthesis should be included on the radiographs. The mar-

This review received Institutional Review Board approval.
[a] Weill Medical College of Cornell University, New York, NY, USA
[b] Hospital for Special Surgery, New York, NY, USA
* Department of Radiology and Imaging, Hospital for Special Surgery, 535 East 70th Street, New York, NY 10021.
E-mail address: sofkac@hss.edu

doi:10.1016/j.rcl.2006.01.008

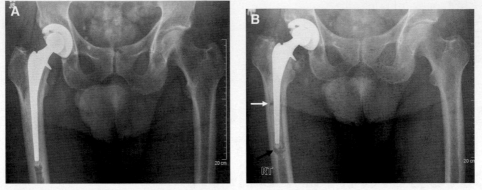

Fig. 1. Usefulness of radiographs for the detection of osteolysis and loosening in a patient who had a painful total hip arthroplasty. (*A*) Early postoperative radiograph taken when the patient was asymptomatic demonstrates a fairly unremarkable appearance of the prosthesis. (*B*) Subsequent radiograph taken years later demonstrates an expansile lucency with endosteal thinning at the femoral tip of the prosthesis consistent with osteolysis and loosening (*black arrow*) and a frank fracture through the cement mantle more superiorly (*white arrow*).

gins of the components of an arthroplasty, especially in the long bones, can act as a stress riser and a fulcrum for the development of a stress, or frankly complete, fracture. Osteolysis also may act to weaken the bone and result in a "pathologic" fracture of either the medullary or cortical bone or cement [Fig. 1]. In other implantable devices, hardware may become dislodged, often clinically unsuspected, and this may be readily apparent on conventional radiographs.

In an initial work-up of a joint or limb postoperatively, especially if there is a documented history of trauma, an initial evaluation should begin with plain film radiographs.

Nuclear medicine scintigraphy

Nuclear medicine scintigraphy still plays an active role in some centers in the evaluation of painful arthroplasty. Patients often present clinically with nonspecific and nonlocalizing pain despite normal alignment of arthroplasty on radiographs.

Nuclear medicine bone scan can indicate areas of abnormal bone turnover, such as those areas replaced by osteolysis or affected by a fracture. The presence of linear uptake at the margins of a prosthesis, especially in long bones, often indicates an area of stress reaction or fracture.

Although nuclear medicine examinations are sensitive to osseous pathology, they often are nonspecific, hindering their diagnostic capabilities. Diagnoses often are a diagnosis of exclusion after other possibilities, such as indolent infection, are excluded from an image-guided arthrocentesis. Nuclear scintigraphic uptake can occur routinely for a moderate amount of time after placement of an arthroplasty, further limiting diagnostic yield. It is

demonstrated that there is persistent radiotracer uptake around the tibial components of a total knee arthroplasty for up to a year after implantation [1,2]. Combined radiotracers or inflammation specific agents, however, help increase the specificity of nuclear scintigraphy for diagnosing infection postoperatively [3].

Although nuclear medicine scintigraphy largely has been replaced by MR imaging, it still has a role in the investigation of global osseous pathology, such as metastatic disease, and in the evaluation of more focal osseous abnormalities, such as regional pain after surgery, in patients who are non–MR compatible because of either absolute or relative contraindications to MR imaging. Nuclear medicine examinations should be interpreted in concert with plain film radiographs and clinical history to increase diagnostic accuracy, and combined radiotracers or inflammation specific agents should be used as indicated to increase diagnostic yield.

Ultrasound

Ultrasound is not limited by regional metallic artifacts. Ultrasound can be used to evaluate painful arthroplasty and potential complications of hardware affecting the regional soft tissues. Sonography is demonstrated as useful in imaging regional soft tissue collections in painful arthroplasty and in guiding for diagnostic arthrocenteses [4].

Ultrasound also has broader applications for painful arthroplasty in that it can evaluate the integrity of the surrounding muscles and tendons, such as the rotator cuff, after total shoulder arthroplasty [5]. The quality of the regional muscles, in addition to the tendons, can be evaluated with sonography demonstrating the presence of potential muscle atrophy [5]. The subscapularis, a tendon that often fails in the setting of a shoulder arthroplasty, usually can be identified clearly with sonography, because it is an anterior, and relatively superficial, structure [5,6]. The application of power Doppler also can help direct radiologists to areas of active inflammation, such as synovitis or tendinitis, thus suggesting areas of potential sonographic-guided therapeutic interventions.

A distinct advantage of sonography over other imaging modalities is its dynamic capabilities. Not only can sonography guide various diagnostic and therapeutic injections, but also sonography can determine the relationship between indwelling implants and the surrounding soft tissues. Because sonography is not limited by regional artifacts, sonography can visualize the integrity of the components of a joint arthroplasty in some settings [5,7]. In addition, although plain film radiographs and static imaging modalities can visualize the relationship of orthopedic hardware and the osseous structures, they cannot always determine the dynamic relationship between the hardware and the surrounding ligaments and tendons [8]. During a dynamic sonographic examination, tendons, such as those about the ankle or hip, often can be shown to abut or move against implantable devices, such as screws or prosthetic implants, and can be correlated with patient symptomatology or diagnostic injections during real-time imaging, such as iliopsoas tendon impingement in total hip arthroplasty [Fig. 2] [9].

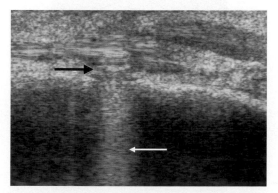

Fig. 2. Sonographic evaluation of soft tissue complications from implantable hardware. Longitudinal sonographic image was obtained along the lateral joint line in a patient who had lateral sided ankle pain and indwelling screws and plates for calcaneal fracture fixation. Note the screw head abuts the deep margin of the peroneus brevis tendon (*black arrow*). During real-time sonographic evaluation, direct irritation of the deep margin of the tendon against the screw head was identified. Note the characteristic reverberation artifact from the metal screw head (*white arrow*).

In summary, sonography has a broad range of applications in the evaluation of patients who have pain after an orthopedic surgical procedure, such as arthroplasty placement. Sonography can be used to evaluate for periprosthetic collections and the integrity of the regional tendons and ligaments. The dynamic capabilities of sonography allow it to be used to guide various therapeutic and diagnostic injections and evaluate the regional relationships between the soft tissue structures and implantable hardware.

CT

CT provides exquisite fine osseous detail in a preoperative setting, demonstrating high usefulness in evaluating fracture fragment alignment, for example. Postoperatively, however, in the presence of (often extensive) metal implants, the diagnostic quality of CT often is reduced if appropriate modifications of CT protocols are not used.

The degree of artifact produced in CT imaging in the presence of hardware is proportional to the amount of metal implanted. In addition, the type of metal implanted affects the degree of artifact, with titanium yielding fewer artifacts than cobalt chrome [10,11].

Parameters that can be modified by radiologists include selecting how the body part is positioned in relation to the CT gantry. Positioning patients such that the x-ray beams course through the smallest diameter or cross-sectional area of the hardware, if possible, is optimal [12]. In addition, increasing the peak kilovoltage (kVp) and milliamperes of tube current (mAs) can improve overall image quality of metal implants [13]. The advent of multidetector scanners has allowed for improved imaging of patients who have metallic implants with the ability to image implants with multiple overlapping slices, resulting in overall increased effective mAs, without resulting in excessive radiation exposure [12].

Postprocessing workstations have allowed for reformatting of images in multiple planes, often resulting in improved visualization of pathology around implanted hardware. The choice of viewing images with either soft tissue or bone windows can improve image interpretation. It is demonstrated that in the presence of bulky hardware, such as a total hip arthroplasty, viewing images with a soft tissue algorithm results in better visualization of regional osseous pathology, such as osteolysis, in contrast to viewing images with bone windows [12]. In contrast, evaluating structures around smaller implantable orthopedic devices, such as small cortical screws, may be afforded better with viewing images in a bone algorithm [12].

The applicability of CT in evaluating postoperative patients is broad and can include the evaluation of radio-occult fractures, guiding diagnostic or therapeutic interventions, or the evaluation of more infiltrative pathology, such as osteolysis in painful arthroplasty [14]. The usefulness of CT in evaluating osteolysis in painful total hip arthroplasty includes its improved ability to visualize areas of osteolysis given its tomographic capabilities, in contrast to conventional radiographs [14].

Postoperative diagnostic CT images can be produced if attention is directed to modifying imaging parameters appropriately. The use of postprocessing workstations to evaluate orthopedic implants in multiple orthogonal planes also can increase diagnostic yield.

MR imaging

Postoperative MR imaging traditionally has been limited, as conventional MR imaging protocols, such as standard spin-echo imaging, often yield nondiagnostic images. The global usefulness of fast spin-echo imaging has been extended to the postoperative setting, with the benefits of fast imaging times, less motion degradation of image quality, and improved diagnostic images [15–20]. Even if there are no bulky metallic implantable devices, but patients have had an orthopedic surgical procedure, such as labral repair with bioabsorbable implants, or are status post cartilage or meniscal repair, pulse sequence parameter modifications are advocated, as even fine metallic shavings from an arthroscope can remain in the knee or shoulder, for example, and can result in bothersome metallic artifacts if not addressed [21,22].

Parameters affecting the amount of magnetic susceptibility generated during an MR examination include the metallic composition of the prosthesis, the morphology of the implants, and the relationship of the implanted material to the main magnetic field [19]. Titanium implants typically result in fewer artifacts than stainless steel [23,24]. The morphology of the implant also has a direct relationship with the total amount of artifact generated. Implants that are fairly uniform in construct and are linear result in fewer artifacts than those that are round or of nonuniform shape [24,25].

The amount of artifact generated by a prosthesis is related to its relationship to the main magnetic field; patients who have implants that can be aligned in a linear fashion, parallel to the main magnetic field, such as an intramedullary nail in the femur, produce fewer artifacts than those who have implants that are nonuniform or spherical in shape. Orienting the frequency-encoding direction along the main longitudinal axis of the implanted

hardware and increasing the frequency-encoding gradient strength are shown to result in visibly decreased artifact [15,26].

Routine pulse sequence parameters often used preoperatively in the evaluation of the musculoskeletal system, such as conventional spin-echo imaging, frequency-selective fat suppression, and gradient-echo imaging, are of limited usefulness postoperatively. With fast spin-echo imaging, short interecho spacing allows less time for dephasing to occur [17,24]. Increasing echo train length plus the inherent short interecho spacing with fast spin-echo imaging decrease chances of mismapping, resulting in less image distortion in contrast to conventional spin-echo imaging [15,17,24]. Increasing the readout bandwidth can help decrease metallic artifact in a postoperative setting, with an inverse relationship between the readout bandwidth and the degree of linear misregistration and secondary gains of shortened interecho spacing [15,19]. Despite an inverse relationship between receiver bandwidth and signal-to-noise ratio, which may be counteracted by the necessary increase in the number of acquisitions in the setting of metal, this is negligible compared with the beneficial effects of decreased chemical shift artifact with increasing receiver bandwidth and contributions to increasing the amplitude of the frequency-encoding gradient [15,19].

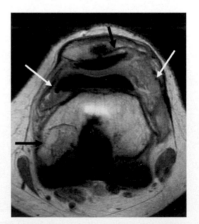

Fig. 3. MR imaging of osteolysis and reactive synovitis in a patient who had a painful total knee arthroplasty. Routine radiographs (not presented) demonstrated satisfactory alignment of the prosthesis. Axial fast spin-echo image demonstrates foci of osteolysis about the femoral component and patellar resurfacing interface as demonstrated by areas of moderately expansile, well marginated, somewhat lobular areas of intermediate signal intensity (*black arrows*). A reactive synovitis also is incited, distending the pseudocapsule (*white arrows*).

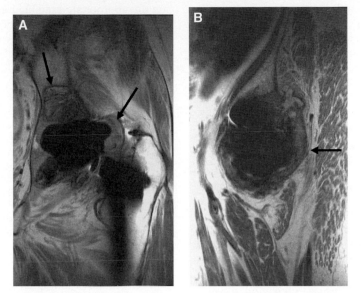

Fig. 4. MR imaging of osteolysis and synovial reaction in a patient who had painful total hip arthroplasty 13 years after it was placed. (*A*) Coronal fast spin-echo image demonstrates marked osteolysis seen as areas of well marginated, slightly expansile, foci of intermediate signal intensity about the acetabular component and distending the pseudocapsule (*arrows*). (*B*) Sagittal fast spin-echo image in the same patient demonstrates moderate distention of the pseudocapsule with debris and synovial reaction, in close proximity to the fascicles of the sciatic nerve (*arrow*).

For the evaluation of musculoskeletal patients, especially those in whom the clinical concern is acute soft tissue or osseous injury, such as a fracture, a water-sensitive pulse sequence, such as frequency-selective fat suppression sequence or fast inversion recovery, typically is used. Postoperatively, a fast inversion recovery sequence is favored in contrast to frequency-selective fat suppression [15,27]. With metallic implants, there is less ability to perceive fat and water molecules as different resonant frequencies, resulting in image distortion and the characteristic "flare" seen about metallic implants with conventional fat suppression [15].

Some investigators advocate the use of gradient-echo imaging, with or without fat suppression, for the evaluation of cartilage pathology [28–31]. These sequences, however, are of limited usefulness postoperatively, as any area of regional magnetic field inhomogeneity results in considerable signal void because of increased intravoxel dephasing (T2* decay) [32,33]. This applies not only to postoperative patients in whom there are metal implants but also to those in whom even minimal arthroscopic shavings are present.

The use of intravenous or intra-articular gadolinium contrast material postoperatively is controversial and highly dependent on radiologist preferences. The universal application of noncontrast MR imaging to evaluate postoperative musculoskeletal patients is demonstrated in a variety of settings, including postoperative cruciate ligament reconstruction, postoperative meniscal and cartilage repair techniques, and total joint arthroplasty [18,20–22,34,35].

Specific clinical applications of MR imaging of postoperative orthopedic patients include, but are not limited to, imaging painful arthroplasty. These patients often have a nonfocal clinical examination and unremarkable alignment of the prosthesis on conventional radiographs. The usefulness of noncontrast fast spin-echo MR imaging in evaluating painful total shoulder, knee, and hip arthroplasty is reported [18,20,34]. With its 3-D tomographic capabilities, MR can demonstrate the presence and extent of osteolysis to better advantage than conventional radiographs, aiding in presurgical planning [Fig. 3]. With its ability to visualize not only the regional osseous structures but also the surrounding soft tissues, potential extraosseous involvement of osteolysis and its relationship to the regional neurovascular structures, also can be discerned [Fig. 4].

Summary

Various methods of imaging postoperative orthopedic patients are available. Plain film radiographs remain the mainstay of diagnostic imaging and should be used to evaluate all patients who have painful joints or hardware after surgery initially. Advanced imaging modalities should be used based on the clinical question posed and on patient factors, such as mobility, fear of closed spaces, and cost restraints. Nuclear medicine provides a sensitive survey evaluation of the skeletal system, indicating potential areas of pathology; however, nuclear scintigraphy using standard radiotracers has limited specificity. Combined radiotracers, such as combined inflammation specific agents, can increase diagnostic yield and specificity.

Sonography plays an increasing role in the evaluation of the musculoskeletal system. Sonography has advantages over other imaging modalities in that it is less expensive, uses no ionizing radiation, and is dynamic. The dynamic capabilities of sonography allow for same-day add-ons, and the ability to perform a diagnostic examination and therapeutic sonography-guided injection at the same time. The dynamic capabilities of sonography also can be applied to evaluate ligament and tendon dynamics and their relationship to adjacent orthopaedic implants.

CT, with proper imaging modifications, can evaluate the soft tissue and osseous structures in postoperative patients. Adjusting the area of interest such that the short axis of the hardware is perpendicular to the x-ray beam, increasing the kVp, and viewing images with various postprocessing software allow for better image quality. CT can evaluate for possible subtle fractures, and, with its tomographic abilities, often can visualize the extent of osteolysis better than conventional radiographs.

MR imaging provides visualization of the osseous structures and the surrounding soft tissues. An advantage of MR over other cross-sectional imaging modalities, such as CT, is that it provides an evaluation of the internal matrix of the bone, identifying areas of subtle bone marrow edema, thus indicating areas of possible pathology, such as a subchondral fracture. Proper modification of imaging parameters must be performed with MR to image postoperative patients appropriately. Fast spin-echo imaging should be used, as the decreased interecho spacing allows for less regional dephasing and resultant decreased metallic susceptibility. Gradient-echo imaging should be avoided in postoperative patients, as even subtle regional field inhomogeneities result in marked signal void. Lastly, as a water-sensitive pulse sequence, fast inversion recovery imaging should be used instead of frequency-selective fat suppression, as it is less susceptible to regional field inhomogeneities.

Knowledge of the basic imaging principles and parameter modifications of the advanced imaging techniques—nuclear medicine, ultrasound, CT, and MR—will aid musculoskeletal radiologists in selecting the appropriate examination for clinical questions posed, and tailoring of the imaging parameters results in better image quality, thus more confident diagnoses.

References

[1] Duus BR, Boeckstyns M, Kjaer L, et al. Radionuclide scanning after total knee replacement: correlation with pain and radiolucent lines, a prospective study. Invest Radiol 1987;22:891–4.

[2] Kantor SG, Schneider R, Insall JN, et al. Radionuclide imaging of asymptomatic versus symptomatic total knee arthroplasties. Clin Orthop Relat Res 1990;260:118–23.

[3] Schneider R, Soudry M. Radiographic and scintigraphic evaluation of total knee arthroplasty. Clin Orthop 1986;205:108–20.

[4] Van Holsbeeck MT, Eyler WR, Sherman LS, et al. Detection of infection in loosened hip prostheses: efficacy of sonography. AJR Am J Roentgenol 1994;163:381–4.

[5] Sofka CM, Adler RS. Sonographic evaluation of shoulder arthroplasty. AJR Am J Roentgenol 2003; 180:1117–20.

[6] Cuomo F, Checroun A. Avoiding pitfalls and complications in total shoulder arthroplasty. Orthop Clin North Am 1998;29:507–18.

[7] Sofka CM, Adler RS, Laskin R. Sonography of polyethylene liners used in total knee arthroplasty. AJR Am J Roentgenol 2003;180:1437–41.

[8] Jacobson JA, Lax MJ. Musculoskeletal sonography of the postoperative orthopedic patient. Semin Musculoskel Radiol 2002;6:67–77.

[9] Wank R, Miller TT, Shapiro JF. Sonographically guided injection of anesthetic for iliopsoas tendinopathy after total hip arthroplasty. J Clin Ultrasound 2004;32:354–7.

[10] Ebraheim NA, Coombs R, Rusin JJ, et al. Reduction of post-operative CT artifacts of pelvic fractures by use of titanium implants. Orthopaedics 1990;13:1357–8.

[11] Haramati N, Staron RB, Mazel-Sperling K, et al. CT scans through metal: scanning technique versus hardware composition. Comput Med Imag Graph 1994;18:429–34.

[12] White LM, Buckwalter KA. Technical considerations: CT and MR imaging in the postoperative orthopedic patient. Semin Musculoskel Radiol 2002;6:5–17.

[13] Robertson DD, Weiss PJ, Fishman EK, et al. Evaluation of CT techniques for reducing artifacts in the presence of metallic orthopedic implants. J Comput Assist Tomog 1988;12:236–41.

[14] Claus AM, Totterman SM, Synchterz CJ, et al. Computed tomography to assess pelvis lysis after total hip replacement. Clin Orthop 2004;422: 167–74.

[15] White LM, Kim JK, Mehta M, et al. Complications of total hip arthroplasty: MR imaging—initial experience. Radiology 2000;215:254–62.

[16] Eustace S, Goldberg R, Williamson D, et al. MR

imaging of soft tissues adjacent to orthopaedic hardware: techniques to minimize susceptibility artifact. Clin Radiol 1997;52:589–94.

[17] Tartaglino LM, Flanders AE, Vinitski S, et al. Metallic artifacts on MR images of the postoperative spine: reduction with fast spin-echo techniques. Radiology 1994;190:565–9.

[18] Sofka CM, Potter HG, Figgie M, et al. Magnetic resonance imaging of total knee arthroplasty. Clin Orthop Relat Res 2003;406:129–35.

[19] Sofka CM, Potter HG. MR imaging of joint arthroplasty. Semin Musculoskel Radiol 2002;6: 79–85.

[20] Potter HG, Nestor BJ, Sofka CM, et al. Magnetic resonance imaging after total hip arthroplasty: evaluation of periprosthetic soft tissue. J Bone Joint Surg [Am] 2004;86-A:947–54.

[21] van Trommel MF, Potter HG, Ernberg LA, et al. The use of noncontrast magnetic resonance imaging in evaluating meniscal repair: comparison with conventional arthrography. Arthroscopy 1998;14:2–8.

[22] Potter HG, Rodeo SA, Wickiewicz TL, et al. MR imaging of meniscal allografts: correlation with clinical and arthroscopic outcomes. Radiology 1996;198:509–14.

[23] Olscamp AJ, Tao SS, Savolaine ER, et al. Postoperative magnetic resonance imaging evaluation of Pipkin fractures fixated with titanium implants: a report of two cases. Am J Orthoped 1997;26:294–7.

[24] Henk CB, Brodner W, Grampp S, et al. The postoperative spine. Top Magn Reson Imaging 1999; 10:247–64.

[25] Mueller PR, Stark DD, Simeone JF, et al. MR-guided aspiration biopsy: needle design and clinical trials. Radiology 1986;161:605–9.

[26] Frazzini VI, Kagetsu NJ, Johnson CE, et al. Internally stabilized spine: optimal choice of frequency-encoding gradient direction during MR imaging minimizes susceptibility artifact from titanium vertebral body screws. Radiology 1997;204:268–72.

[27] Hilfiker P, Zanetti M, Debatin JF, et al. Fast spin-echo inversion-recovery imaging versus fast T2-weighted spin-echo imaging in bone marrow abnormalities. Invest Radiol 1995;30:110–4.

[28] Kornaat PR, Reeder SB, Koo S, et al. MR imaging of articular cartilage at 1.5T and 3.0T: comparison of SPGR and SSFP sequences. Osteoarthritis Cartilage 2005;13:338–44.

[29] Yoshioka H, Stevens K, Hargreaves BA, et al. Magnetic resonance imaging of articular cartilage of the knee: comparison between fat-suppressed three-dimensional SPGR imaging, fat-suppressed FSE imaging, and fat-suppressed three-dimensional DEFT imaging, and correlation with arthroscopy. J Magn Reson Imag 2004;20:857–64.

[30] Yoshioka H, Alley M, Steines D, et al. Imaging of the articular cartilage in osteoarthritis of the knee joint: 3D spatial-spectral spoiled gradient echo vs. fat-suppressed 3D spoiled gradient-echo MR imaging. J Magn Reson Imag 2003;18:66–71.

[31] Cicuttini F, Forbes A, Asbeutah A, et al. Comparison and reproducibility of fast and conventional spoiled gradient-echo magnetic resonance sequences in the determination of knee cartilage volume. J Orthop Res 2000;18:580–4.

[32] Hendrick RE. Basic physics of MR imaging: an introduction. Radiographics 1994;17:829–46.

[33] Taber KH, Herrick RC, Weathers SW, et al. Pitfalls and artifacts encountered in clinical MR imaging of the spine. Radiographics 1998; 18:1499–521.

[34] Sperling JW, Potter HG, Craig EV, et al. MRI of the painful shoulder arthroplasty. J Shoulder Elbow Surg 2002;11:315–21.

[35] Schatz JA, Potter HG, Rodeo SA, et al. MR imaging of anterior cruciate ligament reconstruction. AJR Am J Roentgenol 1997;169:223–8.

RADIOLOGIC
CLINICS
OF NORTH AMERICA

Radiol Clin N Am 44 (2006) 331–341

ELSEVIER
SAUNDERS

Postoperative Imaging of the Shoulder

Kimberly A. Ruzek, MD*, Laura W. Bancroft, MD,
Jeffrey J. Peterson, MD

- Postoperative imaging techniques
- Surgical repair—expected findings and complications
  Impingement
  Rotator cuff tears
  Instability
- Summary
- References

The management of the postoperative shoulder has improved in recent years, in part because of advances in imaging techniques. This is most prominent in MR imaging and MR arthrography where improved coils, software, and sequence techniques have had a major impact on visualization of the postoperative shoulder. As disability and pain may occur or persist after shoulder surgery, it is imperative that clinicians have a detailed anatomic delineation before further treatment. In the past, it often was difficult to distinguish between typical postoperative findings and new pathology. Variations in surgical techniques, surgical distortion of native anatomy, and metallic artifacts decrease the accuracy of postoperative imaging. Pitfalls may be avoided with knowledge of postoperative alterations, such as a non-watertight capsule, irregularity of the rotator cuff tendons, and fluid in the subacromial bursa. This article focuses on normal and abnormal postoperative findings in the shoulder, with emphasis on MR imaging.

Postoperative imaging techniques

In postoperative imaging, it is important to be familiar with the surgical procedure to best evaluate the anatomic structure of concern. In surgical procedures requiring evaluation of osseous components, primary evaluation with conventional radiographs or CT is optimal. If assessment of soft-tissue components is required, MR imaging, including MR arthrography or, occasionally, CT arthrography is best. At the authors' institution a 10- to 12-mL mixture of bupivicane, 2 mmol/L gadopentetate, and iodinated contrast is instilled into the shoulder joint when performing direct MR arthrography. The inclusion of iodinated contrast in the mixture is to "save the case" if a patient becomes unable to tolerate MR imaging. The patient then can undergo CT scanning, which often provides adequate arthrographic imaging of the area of concern.

If recurrent or persistent instability is of concern, MR arthrography is performed for evaluation of

Department of Radiology, Mayo Clinic, Jacksonville, FL, USA
* Corresponding author. Department of Radiology, Mayo Clinic College of Medicine, 4500 San Pablo Boulevard, Jacksonville, FL 32224.
E-mail address: mkruzek@yahoo.com (K.A. Ruzek).

doi:10.1016/j.rcl.2006.02.002

the labroligamentous complex. Some controversy exists as to the role of MR arthrography in postoperative evaluation of rotator cuff pathology. Rates of rerupture after open rotator cuff repair are reported to range from 13% to 68%, depending on the size of the original tear [1,2]. One study finds that patient clinical outcomes with structural failure or rerupture after rotator cuff repair, demonstrated by MR imaging, still had significant improvement in pain and function compared with the preoperative state. The rerupture usually was smaller than the original tear [3]. Subtle changes in the rotator cuff, which may be clinically relevant before surgery, may not be as relevant after surgery. MR athrography is shown to be more accurate in assessment of labroligamentous structures, capsular volume, and the undersurface of the rotator cuff [4,5]. Labral abnormalities are more common and rotator cuff tears are relatively less common in young individuals, in particular athletes [6]. At the authors' institution, the majority of MR arthrograms are performed on young patients or high-performance athletes. Knowledge of specific surgical indications is helpful when choosing to perform MR arthrography in postoperative patients.

MR imaging often can be problematic in the postoperative setting. A dedicated phased-array shoulder coil is optimal. Sequence selection varies between institutions. At the authors' institution, the standard MR shoulder protocol includes an axial gradient and T1-weighted sequences, coronal oblique proton-density and T2-weighted with fat-suppression fast spin-echo (FSE) sequences, and a sagittal proton-density fat-suppressed sequence. An axial T1-weighted sequence is included to assess the degree of muscular fatty degeneration and atrophy for unsuspected marrow abnormalities.

If MR arthrography is indicated, sequence selection is modified. At the authors' institution, standard MR shoulder arthrogram protocol includes: fat-suppressed T1-weighted images in the axial, coronal, and sagittal planes; coronal oblique proton-density and fat-suppressed T2-weighted with FSE sequences; and an additional fat-suppressed T1-weighted sequence with the arm positioned in abduction and external rotation (ABER view). The ABER view is optimal for delineation of the anteroinferior labrum, inferior glenohumeral ligament, and undersurface of the rotator cuff.

Techniques can be used to minimize spatial misregistration as susceptibility artifact with MR imaging in postoperative patients. Unfortunately, techniques that reduce metallic artifacts are inversely related to quality of the signal-to-noise ratio. This can be problematic especially with conventional spin-echo or T2-weighted sequences. Considerations to reduce metallic artifact are dis-

cussed by Sofka elsewhere in this issue and include decreasing field strength, increasing sampling bandwidth, and shortening the echo time [7,8]. Replacement of gradient-echo sequences with FSE sequences also can be useful. Using short tau inversion recovery sequences instead of frequency-selective fat-suppression techniques also can decrease artifacts. Fat-saturation techniques may falter secondary to field inhomogeneity as a result of magnetic susceptibility. Using water excitation instead of spectrally selective fat saturation is another available alternative technique that can reduce artifacts [9].

Surgical repair—expected findings and complications

Imaging commonly is acquired of the postoperative shoulder after surgical intervention for impingement (subacromial decompression), instability (glenohumeral labroligamentous repair), and rotator cuff lesions [10].

Impingement

Subacromial decompression with acromioplasty [Fig. 1] is indicated for signs of extrinsic impingement with an intact rotator cuff. Currently, the procedure is performed most commonly arthroscopically. Surgeon experience with arthroscopic techniques, in addition to patient symptoms and imaging, may warrant consideration of open surgical decompression. Arthroscopically assisted mini-open decompressions and, more recently, completely arthroscopic decompressions increasingly are used. The major disadvantage of open surgical repair is violation of the deltoid attachment and disruption of the overlying fascia. Postoperative dehiscence of

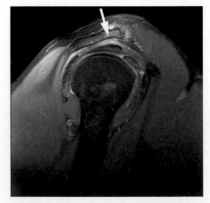

Fig. 1. Acromioplasty. A 45-year-old man underwent rotator cuff repair, subacromial decompression, and acromioplasty several years prior. Sagittal FSE, proton-density, fat-suppressed image depicts the subtle concave inferior contour of the acromion (*arrow*).

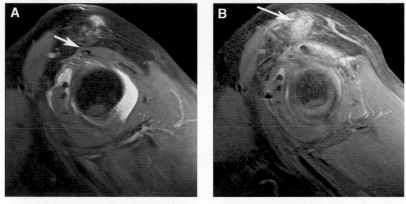

Fig. 2. Clavicular resection in a 76-year-old man. (*A*) Sagittal FSE, proton-density, fat-suppressed image through the acromioclavicular joint demonstrates hypertrophic degenerative change of the joint, with some mass effect on the supraspinatus (*arrow*). (*B*) Sagittal FSE, proton-density, fat-suppressed image after recent distal clavicular resection demonstrates hyperintense soft tissue in the surgical bed (*arrow*) and resolution of the mass effect.

the deltoid repair is reported and results in significant morbidity [11]. Arthroscopic subacromial decompression consists of diagnostic arthroscopy, removal of varying portions of the undersurface and anterior edge of the acromion, including enthesophytes, and, if necessary, distal clavicular resection [Fig. 2]. Usually, during the course of the procedure, it is necessary to remove the acromial attachment of the coracoacromial ligament.

Symptoms that persist after surgical intervention may be unrelated to anatomic impingement. These symptoms still may mimic impingement syndromes clinically as seen in anterior glenohumeral joint instability. Secondarily, persistent or recurrent symptoms may reside from incomplete removal of anatomic compressive structures [12]. It is important in postoperative MR imaging to evaluate the rotator cuff (coexisting tears are another source of

pain), residual acromioclavicular osteoarthritis, and thickening or irregularity of the coracoacromial ligament, if not resected. If the decompressive surgery is too extensive, evaluation for fracture of the acromion, dehiscence of the deltoid muscle, and axillary nerve injury (especially if the deltoid split incision is carried below 5 cm from the level of the acromion) should be considered [13].

Rotator cuff tears

There are many surgical techniques and materials available for shoulder repair. If the applied surgical technique is unknown, imaging evaluation can be difficult. Differing surgical techniques are used if the primary goal is restoration of function or relief from pain [14]. Partial rotator cuff tears respond favorably to débridement. If the partial tears are bursal sided and associated with morphologic

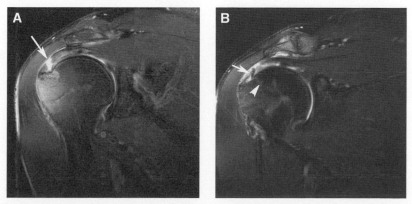

Fig. 3. Repair of supraspinatus tendon tear occurring at the critical zone in a 45-year-old man. Mitek Panalok RC anchors were placed and limbs passed through the supraspinatus tendon and tied down with sliding locking knots and half-stitches. (*A*) Coronal oblique FSE, fat-suppressed, T2-weighted, image through the right shoulder demonstrates a full-thickness tear of the critical zone of the supraspinatus tendon (*arrow*). (*B*) Coronal oblique FSE, fat-suppressed T2-weighted image obtained after repair displays the limb (*arrow*) of the suture anchor (*arrowhead*) passing through the reattached supraspinatus tendon.

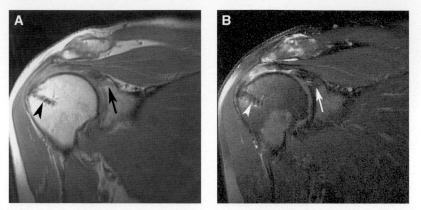

Fig. 4. Rotator cuff and superior labral tear repair in a 62-year-old man. Patient underwent repair of the full-thickness supraspinatus tendon tear with Mytek bioabsorbable corkscrew anchors and superior labral repair with Mytek bioabsorbable anchors. Coronal oblique FSE (*A*) proton-density and (*B*) fat-suppressed T2-weighted images demonstrate the intact bioabsorbable corkscrew rotator cuff (*arrowhead*) and superior labral (*arrowhead*) anchors. Notice the increased signal within the reattached supraspinatus tendon, consistent with granulation tissue.

changes in the coracoacromial arch, then subacromial decompression also is useful. If the partial tears are articular sided, not involving more than two thirds of the tendon, then débridement alone may be satisfactory with or without anterior acromioplasty [15]. When the partial tear is extensive, but not full thickness, the portion of the partially torn tendon may be excised and a full-thickness repair technique applied. Small full-thickness tendon repairs are treated with tendon-to-tendon suturing. In distal full-thickness tears, a tendon-to-bone repair is used [Figs. 3–5] [16]. Occasionally, the greater tuberosity of the humeral head is shaved with an osteotome to promote adherence of the rotator cuff tendons and to facilitate healing.

A variety of surgical tacks and suture material can be used in the repair. Many such surgical tacks are biodegradable and are unapparent on conventional radiographs. MR imaging can demonstrate the remaining bioabsorbable screw and tunnel [Fig. 6]. Osteolysis around the suture anchors may be the result of mechanical forces or focal necrosis resulting from drilling [17]. One study finds that the osseous lucencies can double in size at 6 months; holes around nonabsorbable anchors stabilize, whereas the holes around the absorbable anchors are replaced with bone at 2 years [18].

A variety of surgical techniques also can be used in the rotator cuff repair. For example, as the role of the long head of the biceps tendon becomes

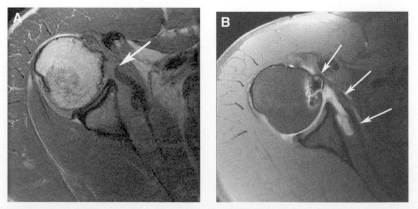

Fig. 5. Repair of subscapularis tendon avulsion in a 53-year-old man. Patient underwent repair of the subscapularis rupture with horizontal sutures and Arthrex suture anchors 7 years prior. (*A*) Axial proton-density image displays the avulsion of the subscapularis tendon from the lesser tuberosity, with retraction of the serpentine tendinous remnant (*arrow*) to the level of the glenohumeral joint. (*B*) Postoperative axial MR arthrogram displays the reattached subscapularis tendon (*arrows*), susceptibility artifact at the suture anchor, and interval mild subscapularis muscle atrophy.

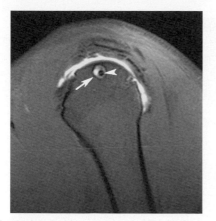

Fig. 6. Bioabsorbable screw and tunnel after rotator cuff repair in a 41-year-old woman who had persistent postoperative pain. Sagittal MR arthrogram demonstrates the intact screw (*arrowhead*) and surrounding tunnel (*arrow*). Imaging and postoperative diagnostic arthroscopy confirmed an intact rotator cuff repair.

increasingly recognized as a component of rotator cuff pathology, indications for biceps tenodesis [Fig. 7] have expanded. The decision for tenotomy or tenodesis may be recommended for irreversible structural changes in the tendon, such as significant atrophy or hypertrophy, partial tearing greater than 25% of the width of the tendon, any subluxation of the tendon from the groove, and for certain disorders of the biceps origin [19]. Tenodesis is preferred for younger patients [19].

After rotator cuff repair, it is common to have either intermediate (granulation tissue) or low signal intensity (fibrotic tissue) in the repaired tendons [Fig. 4]. One article cites that only 10% of repaired tendons demonstrate normal MR imaging (study of 15 patients) [20].

In addition, common ancillary markers associated with rotator cuff tears, such as subacromial-subdeltoid bursal fluid and obliteration of the subacromial-subdeltoid fat in the presurgical population, are not accurate indications of retear in the postsurgical population [21]. The subdeltoid fat is expected to be abnormal postoperatively, as this area is transversed during surgery and often resected. A non-watertight seal is a common postoperative finding and does not correlate with full- or partial-thickness tears. Contrast may leak through a well-repaired but incompletely healed tendon or even through a portal tract [Fig. 8] [20].

Fluid injected from the MR arthrogram also may extend into the acromioclavicular joint. On preoperative imaging, this is referred as the geyser sign and is indicative of a full-thickness tear. In a postoperative shoulder, the undersurface of the acromioclavicular joint commonly is disrupted, thus the geyser sign has no associated implications in the postoperative setting.

Complications do occur after rotator cuff repair. In a Mayo Clinic study of 116 shoulders, there was a 38% overall complication rate and a 16% major complication rate [22]. Complications included frozen shoulder, deep infection, and dislocation in order of decreasing frequency [22]. Other complications of rotator cuff repair included full-thickness retear [Fig. 9], heterotopic ossification [Fig. 10], synovitis, and hardware displacement [2,23–28]. Rotator cuff retear is diagnosed in the same fashion on preoperative imaging with fluid-like signal intensity extending through the substance of the

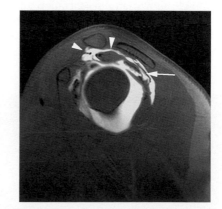

Fig. 8. Contrast extravasation through prior posterior arthroscopy portal, with intact rotator cuff. A 41-year-old woman underwent arthroscopic repair of a type IV superior labrum anterior to posterior tear of the labrum 7 months prior. Sagittal MR arthrogram demonstrates extension of contrast from the glenohumeral joint, through the infraspinatus at site of prior arthroscopic portal (*arrow*), and into the subacromial/subdeltoid bursa (*arrowheads*). Subsequent arthroscopy confirmed an intact rotator cuff.

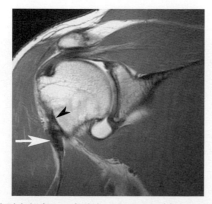

Fig. 7. Bicipital tenodesis in a 31-year-old man. Coronal T1-weighted MR arthrogram demonstrates minimal artifact from the bicipital tenodesis (*arrow*), intact biceps attachment (*arrowhead*), and irregularity of the superior labrum. Omitting fat suppression improved susceptibility artifact about the tenodesis.

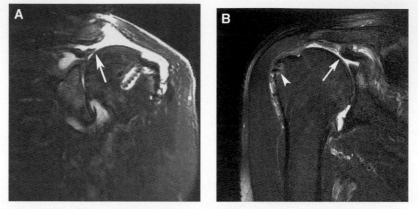

Fig. 9. Supraspinatus tendon retear. (*A*) Coronal oblique FSE, fat-suppressed, T2-weighted image in an 80-year-old woman demonstrates full-thickness supraspinatus retear (*arrow*), intact bioabsorbable corkscrew anchor, large amount of joint fluid, narrowing and remodeling of the glenohumeral articulation, and atrophy of the visualized rotator cuff musculature. Note the hyperintense signal in the tunnel surrounding the screw. (*B*) Coronal oblique, fat-suppressed, T2-weighted image in a different patient demonstrates similar findings in the right shoulder. The anchor (*arrowhead*) remains intact; however, there is a supraspinatus tendon retear (*arrow*).

tendon or nonvisualization of a portion of the tendon [Fig. 5] [24]. The MR diagnosis of partial-thickness tears is more controversial. Partial-thickness tears often are indistinguishable from intact repaired tendons; the visualization of irregularity in morphology and signal can be normal [24]. Rotator cuff failure can have a variety of causes—suture-bone or suture-anchor pullout, suture breakage, knot slippage, tendon pullout, poor quality tendon or bone, muscle atrophy, or inadequate initial repair or improper physical therapy. The likelihood of a recurrent tear is much greater for tendons whose muscle shows fatty degeneration and atrophy as described originally by Goutallier and colleagues in CT imaging and confirmed by Fuchs and coworkers in MR imaging [29–31]. It is important

on postoperative imaging to note progression or regression of muscle atrophy, as this has important prognostic implications [Fig. 11].

Instability

Complex factors, including anatomic development, contribute to shoulder instability; however, the most common causative factor is trauma. This is a common condition particularly in young patients and athletes. The most common instability is anterior glenohumeral and correlates with capsular and/or labroligamentous deficiencies and often is after anteroinferior dislocation [32]. Although there is a plethora of potential sites for traumatic deficiencies [Fig. 12], two general categories of surgical techniques are used: direct anatomic repair

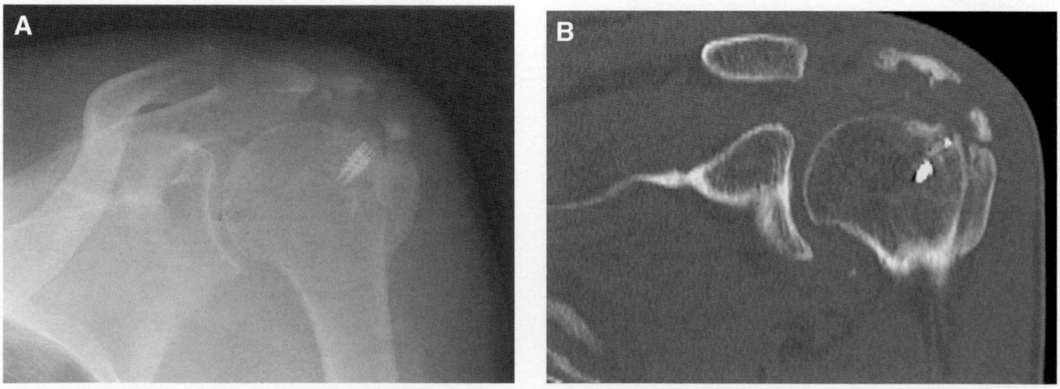

Fig. 10. Extensive heterotopic ossification in a 70-year-old man 6 months after rotator cuff repair. (*A*) AP radiograph and (*B*) coronal CT reconstructed image demonstrate mature heterotopic ossification within the substance of deltoid surrounding the lateral humeral head. Rotator cuff metallic tendon suture anchors are well positioned in the humeral head.

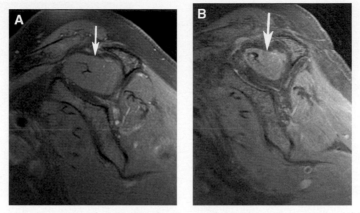

Fig. 11. Progressive supraspinatus atrophy associated with partial-thickness tendon retear. (*A*) Sagittal FSE, proton-density, fat-suppressed image obtained before arthroscopic repair of partial-thickness supraspinatus tendon tear demonstrates mild muscular atrophy (*arrow*). (*B*) Sagittal FSE, proton-density, fat-suppressed image obtained 11 months later now displays further atrophy (*arrow*).

of the glenoid/labrum/capsule complex (Bankart repair) and nonanatomic reconstructions preventing secondary dislocation. Anatomic repairs do not alter the native anatomy of the shoulder and are performed more commonly.

Direct anatomic repair involves suturing the anterior labrum and joint capsule, usually including the anterior band of the inferior glenohumeral ligament, to the glenoid rim. After anatomic procedures there should be no separation of the labroligamentous complex and the glenoid margin. Nonanatomic reconstructions can involve soft tissue or osseous structures. If the anterior capsule is insufficient, the capsule can be cut in a "T" fashion and the free flaps overlapped. This results in capsular tightening and overall reduction in the anteroinferior capsular volume [16]. This is the capsular shift procedure and is used as a primary treatment

for multidirectional instability or as an addition to the classic Bankart repair if patients have pronounced laxity [Fig 13]. Lasers and thermal energy are used arthroscopically to shrink the capsular tissue to try to replicate the morphology of the capsular shift procedure without the considerations for open repair required by the classical capsular shift. Recently, however, these have fallen out of favor because of suboptimal results. Bone block osteotomies are not used commonly except in reconstructions in revision surgery. Bone usually is derived from the coracoid process and grafted to the anteroinferior glenoid rim to block recurrent dislocation [15].

Common expected findings after surgical repairs for glenohumeral instability can include mild superior subluxation of the humeral head, which may be caused by capsular tightening, scarring, or

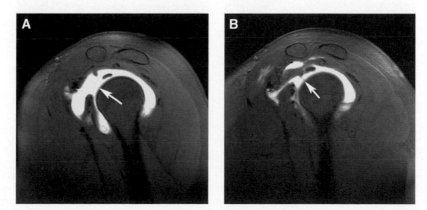

Fig. 12. Rotator interval repair in a 30-year-old woman. (*A*) Sagittal MR arthrogram through the left shoulder demonstrates a large, gaping rotator interval (*arrow*). (*B*) Sagittal MR arthrogram obtained 2 months after side-to-side repair of the rotator interval demonstrates a less capacious interval (*arrow*).

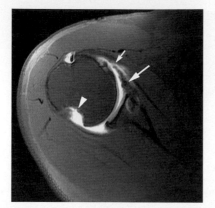

Fig. 13. Failed anterior capsulorrhaphy and Hill-Sachs lesion. A 43-year-old man underwent anterior capsular repair for history of instability and recurrent shoulder dislocations. Axial MR arthrogram demonstrates avulsion (*small arrow*) of the anterior capsule, which is thickened and irregular (*large arrow*) from prior capsulorrhaphy. Note the large Hill-Sachs lesion (*arrowhead*) from repeated dislocations and persistent instability.

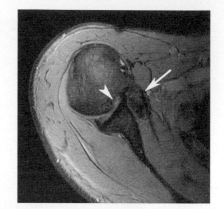

Fig. 15. Anterior capsulorrhaphy. A 39-year-old man previously underwent capsular shrinkage procedure for instability. Axial gradient MR image demonstrates marked thickening of the low-signal anterior capsule and supraspinatus tendon (*arrow*), with persistent anterior subluxation of the humeral head. Note impaction injury of the humerus (*arrowhead*) from prior dislocation.

bursectomy. Expected MR findings after labral repair include susceptibility artifacts along the anterior glenoid if metallic anchors are used or anchor tracks if bioabsorbable tacks are used [see **Fig. 4**; **Fig. 14**]. Another common finding after repair for instability is capsular thickening but, more specifically, an irregular nodular capsular appearance [**Fig. 15**] [33,34]. Postoperative evaluation of the repaired labral or Bankart lesions may be difficult, as granulation tissue processes may occur, limiting demarcation by joint fluid or a contrast agent; this

in itself may be indicative of a successful surgical result. As with all radiographic imaging, comparison with prior studies is imperative for optimal interpretation.

Postoperative complications for glenohumeral instability or labral tears include recurrence of instability and retear [**Fig. 16**]. Tears of the capsule can occur with disruption of the postoperatively thickened capsule and associated secondary anteroinferior subluxation of the humeral head [**Fig. 7**].

In addition to recurrence of prior pathology, it is important to describe migration of hardware, spe-

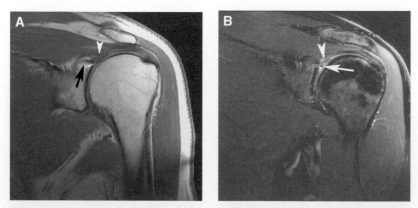

Fig. 14. Intact labral repair in a 55-year-old man. Patient underwent posterosuperior labral repair with Mitek G2 anchors 16 months prior. Coronal oblique FSE (*A*) proton-density and (*B*) T2-weighted fat-suppressed images demonstrate a residual increased signal with the superior labrum (*arrowhead*) and an intact superior labral anchor. Note the screw head extends beyond the cortical margin and abuts the humeral head cartilage (*arrow*). Repeat arthroscopy confirmed intact superior labrum and anchors.

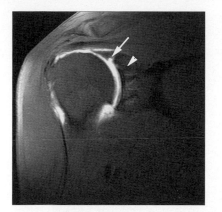

Fig. 16. Failed superior labral repair in a 43-year-old man. Coronal oblique MR arthrogram demonstrates a recurrent superior labral tear (*arrow*) and remaining labral anchor (*arrowhead*).

cifically of surgical bioabsorbable tacks or suture anchors. These may protrude [Fig. 14] or become completely loose and contact the articular cartilage leading to early degeneration. Failure or premature degeneration of bioabsorbable tacks can lead to dislodgment and creation of an iatrogenic loose body within the shoulder [Fig. 17]. Unfortunately, bioabsorbable tacks are not seen on conventional radiographs. Postoperative MR imaging can demonstrate the tacks and their tunnels for delineation of dislodgement and can evaluate for associated synovitis, which can be induced by the bioabsorbable tacks [Fig. 8] [35].

Erroneous placement of the tacks or suture anchors can cause nerve injury. The axillary nerve crosses the inferolateral surface of the subscapularis muscle and the anteroinferior glenohumeral joint capsule, rendering it susceptible to injury in surgery for glenohumeral instability [15]. If a surgical anchor is placed posteriorly, it can impinge on

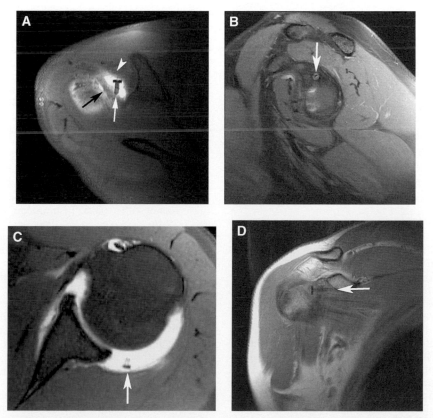

Fig. 17. Loose biodegradable anchors. (*A*) Axial MR arthrogram through the superior shoulder demonstrates a fractured, dislodged bioabsorbable labral anchor in the rotator interval (*white arrow*) between the superior glenohumeral ligament (*arrowhead*) and biceps tendon (*black arrow*). (*B*) Sagittal FSE, proton-density, fat-suppressed image in the same patient demonstrates a portion of the anchor (*arrow*) remaining within the superior glenoid tunnel (*arrow*). (*C*) Axial MR arthrogram in a different patient demonstrates a loose biodegradable screw head fragment (*arrow*) in the posterior joint space. (*D*) Coronal oblique FSE, proton-density image in another patient demonstrates a dislodged labral anchor (*arrow*) in the rotator interval.

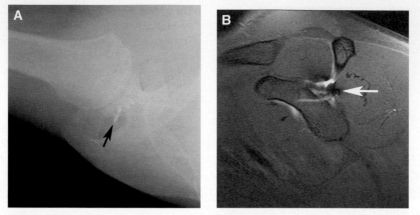

Fig. 18. Suboptimal labral suture anchor in the spinoglenoid notch in a 44-year-old man. (*A*) Radiograph in the axillary projection demonstrates one of three labral anchors extending posteriorly into the spinoglenoid notch (arrow). (*B*) Sagittal MR arthrogram demonstrates suboptimal screw placement into the spinoglenoid notch (arrow). Despite patient symptoms, no muscular atrophy was evident.

the spinoglenoid notch and injure the suprascapular nerve [Fig. 18].

Summary

In conclusion, knowledge of preoperative anatomic delineation, surgical techniques, and expected postoperative anatomy is essential for the accurate interpretation of shoulder imaging. Successful and failed attempts at postsurgical repair of shoulder pathology are presented, including those of the rotator cuff, labrum, and capsule. A thorough understanding of the postoperative shoulder enables radiologists to guide imaging, assess examinations accurately, and assist orthopedic surgeons in proper treatment of patients.

References

[1] Bellumore Y, Mansat M, Assoun J. Results of the surgical repair of the rotator cuff. Radio-clinical correlation. Rev Chir Orthop 1994;80:582–94.

[2] Harryman DT, Mack LA, Wang KY, et al. Repairs of the rotator cuff. Correlation with functional results with integrity of the cuff. J Bone Joint Surg [Am] 1991;73-A:982–9.

[3] Jost B, Pfirrmann CW, Gerber C. Clinical outcome after structural failure of rotator cuff repairs. J Bone Joint Surg [Am] 2000;82-A:304–14.

[4] Beltran J, Bencardino J, Mellando JM, et al. MR arthrography of the shoulder: normal variants and pitfalls. Radiographics 1997;17:1403–12.

[5] Stoller DW. MR arthrography of the glenohumeral joint. Radiol Clin North Am 1997;35:97–115.

[6] Magee T, Williams D, Mani N. Shoulder MR arthrography: which patient group benefits most? AJR Am J Roentgenol 2004;183:969–74.

[7] Parizel PM, van Hasselt BA, van den Hauwe L, et al. Understanding chemical shift induced boundary artifacts as a function of field strength: influence of imaging parameters (bandwidth, field-of-view, and matrix size). Eur J Radiol 1994;18:158–64.

[8] Viano AM, Gronemeyer SA, Haliloglu M, et al. Improved MR imaging for patients with metallic implants. Magn Reson Imaging 2000;18:287–95.

[9] Hauger O, Dumont E, Chateil JF, et al. Water excitation as an alternative to fat saturation in MR imaging: preliminary results in musculoskeletal imaging. Radiology 2002;224:657–63.

[10] Resnick D. Shoulder. In: Resnick D, Kang HS, editors. Internal derangement of joints: emphasis on MR imaging. Philadelphia: Saunders; 1997. p. 163–333.

[11] Norberg FB, Field LD, Savoie 3rd FH. Repair of the rotator cuff. Mini-open and arthroscopic repairs. Clin Sports Med 2000;19:77–99.

[12] Cordasco FA, Bigliani LU. The treatment of failed rotator cuff repairs. Instr Course Lect 1998;47:77–86.

[13] Beals TC, Harryman 2nd DT, Lazarus MD. Useful boundaries of the subacromial bursa. Arthroscopy 1998;14:465–70.

[14] Zlatkin MB. MRI of the postoperative shoulder. Skeletal Radiol 2002;31:63–80.

[15] Mohana-Borges AV, Chung CB, Resnick D. MR imaging and MR arthrography of the postoperative shoulder: spectrum of normal and abnormal findings. Radiographics 2004;24:69–85.

[16] Zanetti M, Hodler J. MR imaging of the shoulder after surgery. Magn Reson Imaging Clin North Am 2004;12:169–83.

[17] Glueck D, Wilson T, Johnson DL. Extensive osteolysis after rotator cuff repair with a bioabsorbable suture abchor: a case report. Am J Sports Med 2005;33:742–4.

[18] Warme WJ, Arciero RA, Savoie FH, et al. Nonabsorbable versus absorbable suture anchors for open Bankart repair. Am J Sports Med 1999;27:743–6.

[19] Ball C, Galatz LM, Yamaguchi K. Tenodesis or

tenotomy of the biceps tendon: why and when to do it. Techn Shoulder Elbow Surg 2001;2:140–52.

[20] Spielmann AL, Forster BB, Kokan P, et al. Shoulder after rotator cuff repair: MR imaging findings in asymptomatic individuals—initial experience. Radiology 1999;213:705–8.

[21] Needell SD, Zlatkin MB, Sher JS, et al. MR imaging of the rotator cuff: peritendinous and bone abnormalities in an asymptomatic population. AJR Am J Roentgenol 1996;166:863–7.

[22] Mansat P, Cofield RH, Kersten TE, et al. Complications of rotator cuff repair. Orthop Clin North Am 1997;28:205–13.

[23] Cummins CA, Murrell GA. Mode of failure for rotator cuff repair with suture anchors identified at revision surgery. J Shoulder Elbow Surg 2003; 12:128–33.

[24] Owen RS, Iannotti JP, Kneeland JB, et al. Shoulder after surgery: MR imaging with surgical validation. Radiology 1993;186:443–7.

[25] Kelly JD. Disintegration of an absorbable rotator cuff anchor six weeks after implantation. Arthroscopy 2005;21:495–7.

[26] Lee S, Mahar A, Bynum K, et al. Biomechanical comparison of bioabsorbable sutureless screw anchor versus suture anchor fixation for rotator cuff repair. Arthroscopy 2005;21:43–7.

[27] Magee T, Shapiro M, Hewell G, et al. Complications of rotator cuff surgery in which bioabsorbable anchors are used. AJR Am J Roentgenol 2003;181:1227–31.

[28] Cummins CA, Strickland S, Appleyard RC, et al. Rotator cuff repair with bioabsorbable screws: an in vivo and ex vivo investigation. Arthroscopy 2003;19:239–48.

[29] Goutallier D, Postel JM, Bernageau J, et al. Fatty muscle degeneration in cuff ruptures. Pre- and postoperative evaluation by CT scan. Clin Orthop 1994;304:78–83.

[30] Fuchs B, Weishaupt D, Zanetti M, et al. Fatty degeneration of the muscles of the rotator cuff: assessment by computed tomography versus magnetic resonance imaging. J Shoulder Elbow Surg 1999;8:599–605.

[31] Goutallier D, Postel JM, Gleyze P, et al. Influence of cuff muscle fatty degeneration on anatomic and functional outcomes after simple suture of full-thickness tears. J Shoulder Elbow Surg 2003; 12:550–4.

[32] Beltran J, Rosenberg ZS, Chandnani VP, et al. Glenohumeral instability: evaluation with MR arthrography. Radiographics 1997;17:657–73.

[33] Vahlensieck M, Lang P, Wagner U, et al. Shoulder MRI after surgical treatment of instability. Eur J Radiol 1999;30:2–4.

[34] Rand T, Freilinger W, Breitenseher M, et al. Magnetic resonance arthrography (MRA) in the postoperative shoulder. Magn Reson Imaging 1999; 17:843–50.

[35] Major NM, Banks MC. MR imaging of complications of loose surgical tacks in the shoulder. AJR Am J Roentgenol 2003;180:377–80.

RADIOLOGIC
CLINICS
OF NORTH AMERICA

Radiol Clin N Am 44 (2006) 343–365

Postoperative Imaging of the Hip

Douglas P. Beall, MD[a,b,*], Hal D. Martin, DO[c],
Justin Q. Ly, MD[d], Scot E. Campbell, MD[d],
Suzanne Anderson, MD[e], Moritz Tannast, MD[f]

- Cross-sectional Imaging of the hip
  MR arthrography
  MR imaging without contrast
- Multidetector CT evaluation
- Postoperative imaging evaluation
  Acetabular labral repair
- Femoral head-neck junction osteoplasties
  Open surgical technique
  Arthroscopic technique
  Hip fracture repair
  Core decompression

- Total hip arthroplasty
- Imaging of total hip arthroplasty
- Common pathology in total hip arthroplasty
  Soft tissue abnormalities
  Loosening
- Infection
- Particle inclusion disease
- Summary
- References

Postoperative imaging of the hip has received much less attention than postoperative imaging of the knee or shoulder but is equally important if persistent pathology, surgical complications, or recurrent disease is to be detected. Even preoperatively, the hip is not recognized as the source of symptoms in 60% of patients, and the average length of time from onset of symptoms to diagnosis is approximately 7 months [1]. Postoperative imaging of the hip often requires the use of multiple modalities. There are many surgical procedures that may require imaging follow-up, including acetabular labral repair, head-neck junction osteoplas-

ties, hip fracture repair, core decompression (CD), and total hip arthroplasty (THA) [2–4] [Table 1].

The cross-sectional imaging modalities are among the most useful techniques in the evaluation a postoperative hip. Possibly the most useful modality in this armamentarium of the hip is magnetic resonance (MR) arthrography [Table 1]. This imaging technique involves the injection of saline mixed with a small amount (0.1–0.2 mL) of gadolinium (ie, gadopentetate dimeglumine) into the hip joint via a small-bore needle. This method, known as direct MR arthrography, is most useful in the evaluation of the acetabular labrum and of the capsu-

a Department of Radiology, Oklahoma Sports Science & Orthopaedics, 6205 North Santa Fe, Suite 200, Oklahoma City, OK 73118, USA
b University of Oklahoma, 610 NW 14, Oklahoma City, OK 73101, USA
c Oklahoma Sports Science & Orthopaedics, 6205 North Santa Fe, Suite 200, Oklahoma City, OK 73118, USA
d Department of Radiology and Nuclear Medicine, Wilford Hall Medical Center, 2200 Bergquist Drive, Suite 1, Lackland AFB, TX 78236-5300, USA
e Musculoskeletal Radiology, Department of Diagnostic, Interventional and Pediatric, University Hospital of Bern Switzerland, Friburg Str. CH-3005, 3010 Bern, Switzerland
f Department of Orthopaedic Surgery, University of Bern, Inselspital, Murtenstrasse, 3010 Bern, Switzerland
* Corresponding author. Department of Radiology, Oklahoma Sports Science & Orthopaedics, 6205 North Santa Fe, Suite 200, Oklahoma City, OK 73118.
E-mail address: dpb@okss.com (D.P. Beall).

doi:10.1016/j.rcl.2006.01.003
radiologic.theclinics.com

Table 1: Protocol for MR imaging of the hip: gadolinium arthrogram

Primary sequences	Plane	Sequence	Coil	FOV	Slice	Matrix Nex	TR	TE	Flip angle	Band	ETL	Fat supp?	Other
Localizer	Triplanar	GRE	Torso	40	7/2	288/128 2	175	MIN-FULL	20	16	—	N	2D, GRE, S/I AUTO
1	Axial	T1SE	Torso	20	4/0.5	288/192	400	MIN-FULL	—	16	—	Y	2D,SE,NPW,SAT:S,I, 12 SLICES, R/I AUTO
2	Coronal	T2 FSE	Torso	20	4/0.5	288X256 4	3425	34	—	16	12	Y	2D,SE,NPW,FC,FAST15 SLICES, S/I,PC AUTO FREQ
3	Coronal	T1 SE	Torso	20	4/0.5	288X192 2	400	MIN-FULL	—	16	—	Y	2D,NPW,SAT:S,I, 15 SLICES, S/I AUTO
4	Sagittal	T1 SE	Torso	20	4/0.5	288X192 2	400	MIN-FULL	—	16	—	Y	NPW, 14 SLICES, A/P AUTO
5	Oblique Axial	T1SE	Torso	20	4/0.5	288/192 2	400	MIN-FULL	—	16	—	Y	2D,SE,NPW,SAT:S,I, 12 SLICES
6	Oblique Sagittal	T1 SE	Torso	20	4/0.5	288X192 2	400	MIN-FULL	—	16	—	N	NPW, 14 SLICES, A/P AUTO
Localizer— knees	Axial or triplanar	GRE	Body	40	7/2	288/128 2	175	MIN-FULL	70	64	—	N	2D,GRE, S/I AUTO

Special instructions:

Sequence #1: both hips and surrounding soft tissues in view.

Sequence #2–7: small FOV over hip of interest.

Sequence #6: oblique axial plane oriented parallel to the femoral neck. Coverage is from superior portion of the greater trochanter to the inferior portion of the lesser trochanter.

Sequence #7: oblique sagittal plane oriented perpendicular to the femoral neck. Coverage is from femoral head to intertrochanteric line.

Localizer #2: done through the knees using the body coil.

Note: orient sequences 6 & 7 parallel and perpendicular to the orientation of the femoral neck.

Employ respiratory compensation if breathing artifact is present.

Abbreviations: 2D, two-dimensional; A/P, anterior to posterior; ETL, echo train length; FC, flow compensation; FSE, fast spin echo; GRE, gradient echo; I, inferior; Min-Full, range of minimum to full TEs; NPW, no phase wrap; S, superior; Sat, saturation band; SE, spin echo; S/I, superior to inferior; Supp, suppressed; TE, echo time; TR, repetition time.

loligamentous structures, and for the determination of the presence of loose bodies [5]. The most important advantage of direct MR arthrography is the ability to achieve adequate capsular distension, which allows the intra-articular anatomy to be delineated optimally by the contrast mixture that is injected into the hip joint.

MR arthrography increasingly is used to evaluate the hip, especially in young patients who have hip pain and in patients who have had prior hip surgery. MR imaging is limited by the magnetic susceptibility artifact produced by metal but is used increasingly in postoperative situations, including after THA [6]. Although most investigators advocate the use of direct arthrography when evaluating a hip joint (especially to detect acetabular labral and articular cartilage abnormalities), noncontrast MR imaging of a hip is useful diagnostically, is noninvasive, and, if optimized for detection of intra-articular pathology, may be comparable diagnostically to direct MR arthrography [7].

Multidetector CT (MDCT) also is useful for evaluating the anatomy of the hip and pelvis, including the osseous structures and surrounding soft tissue. The strength of MDCT relates to its multiplanar isotropic scanning capability, its ability to display the osseous structures optimally, and its relative resistance to metallic streak artifact compared with single-slice CT.

Conventional radiography remains an important modality for the evaluation of pre- and postoperative hips. The global view of postsurgical constructs, such as THAs or lag screws and side plates, is useful for assessing component breakage or displacement, and the ability of conventional radiography to demonstrate soft tissue calcifications and various osseous abnormalities is optimal [8,9].

Radionuclide imaging is helpful for detecting a variety of postoperative complications, including infection (soft tissue infection and osteomyelitis), periprosthetic fracture, and loosening of a THA. Although these pathologic processes may be evaluated well with other modalities, radionuclide imaging is helpful especially in cases of THA loosening that may be equivocal by other forms of imaging [10].

Surgical procedures in and around hip joints increased during the past decade and the annual number of THAs is in excess of 800,000, with more than 120,000 performed in the United States [11,12]. The incidence of hip surgery, including hip arthroscopy, also is increasing, and techniques of acetabular labral repair, joint capsule plication, and femoral head-neck junction osteoplasty are more prevalent [13,14]. This article focuses primarily on imaging of the postoperative hip, the most common types of procedures performed,

and the scenarios encountered most commonly after these procedures.

Cross-sectional Imaging of the hip

MR arthrography

The joint distension achieved with direct MR arthrography allows for optimal evaluation of the intra-articular structures of the hip. The hip joint is accessed by placing the lower extremity in neutral rotation and targeting the subcapital region of the femoral neck [Fig. 1]. The primary indications for MR arthrography are evaluation of the capsuloligamentous structures, the acetabular labrum, and intra-articular loose bodies. Direct arthrography provides an optimal evaluation of the integrity of the repaired acetabular labrum and the status of the articular cartilage [13].

Although direct MR arthrography is an optimal imaging technique for the indications described previously, by definition, this procedure is invasive. It also is more expensive than routine MR imaging of the hip and should be reserved for situations in which it has clear usefulness. Additionally, patients are exposed to ionizing radiation during the fluoroscopic placement of the needle, and the examination requires coordinating the fluoroscopy and MR imaging schedules to allow an effective transition of patients to the MR imaging unit after the hip injection. The fluoroscopic suite also must be located near the MR imaging unit to accomplish this study logistically.

The pitfalls of hip joint injection are not as common as with the injection of some other joints. The needle tip must be present within the joint itself rather than in another potential space, such as the iliopsoas bursa; intra-articular confirmation may be obtained with the injection of 1 to 2 μL of iodinated contrast material. Care must be taken in

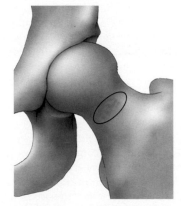

Fig. 1. Diagrammatic representation of the proximal femur with the target location (*oval*) for accessing the hip joint.

order not to overdistend the hip joint, as this can result in extra-articular contrast leaks. Additionally, the injection of air into the joint also occasionally may lead to diagnostic confusion, because air bubbles can mimic loose bodies within the joint. The air bubbles, however, are antidependent in their position within the joint, whereas loose bodies usually are located within the dependent portion of the joint. Other ways to differentiate air from loose bodies is that air bubbles tend to be smaller and have an associated magnetic susceptibility artifact. If the injected gadolinium is not diluted appropriately, a susceptibility artifact also can arise from the injected contrast material. An inadequately diluted gadolinium solution appears as low signal intensity material within the joint. This low signal tends to obscure the surrounding anatomic structures and can limit the overall anatomic assessment severely.

MR imaging without contrast

The majority of the literature supports the use of direct MR arthrography for the evaluation of the acetabular labrum and articular cartilage. Many investigators disparage the use of noncontrast MR imaging and cite sensitivities and specificities as low as 30% and 36%, respectively, versus values in the 90% and higher ranges for MR arthrography [15–21]. Most of the nonarthrographic studies, however, have not optimized the MR imaging sequences and do not compare direct arthrography and noncontrast MR imaging to the pathology found at arthroscopy. Some of the pitfalls in detecting labral tears with MR arthrography apply to noncontrast MR imaging, including the undercutting of articular cartilage at the base of the acetabular labrum and the normal variation of labral shape found with normal aging.

An article by Mintz and colleagues [72], containing data from 1997 to 2000, uses noncontrast MR imaging to evaluate the acetabular labrum and articular cartilage in 92 patients and compares it to the pathology detected at hip arthroscopy. The readers detected 94% to 95% of the tears on noncontrast MR imaging that were detected at hip arthroscopy and graded correctly (within 1 grade) 86% to 92% of the articular cartilage lesions seen in the femoral head and acetabulum. These results were comparable to those reported by MR arthrography and were significantly more accurate than previously reported noncontrast studies [20].

Some of the differences between the information reported by Mintz and most other reports examining the effectiveness of noncontrast MR imaging of the hip is that their MR imaging sequence used was optimized for evaluation of the labrum and articular cartilage. They also used surface coils to attain the highest signal-to-noise ratio (SNR) possible while maintaining a small field of view (FOV). The MR imaging was performed with matrices ranging from 512 × 256 to 512 × 384 and an FOV of 15 to 17 cm for an in-plane resolution of up to 330 to 442 μm. A fast spin-echo proton density sequence was used to maximize the signal difference between synovial fluid, articular cartilage, and underlying bone, and the high-resolution sequence was performed in three planes.

Although most investigators advocate the use of intra-articular contrast when attempting to detect damage to the acetabular labrum or the articular cartilage, the use of noncontrast MR imaging with parameters optimized for the hip joint seems to have the potential to be as effective as MR arthrography without the invasive component. Additional studies are necessary to determine if this is the case and to show that the results can be replicated at other institutions. Mintz and colleagues provide compelling evidence that diagnostically accurate MR imaging of the hip can be performed without intra-articular contrast.

Multidetector CT evaluation

MDCT can provide a high-resolution, multiplanar examination of postoperative hips. The isotropic voxels allow patients' anatomy to be viewed in any plane and MDCT is far less susceptible to the adverse effects of metallic artifact than single-slice CT. The hip may be scanned from the superior portion of the iliac crest to the tibial tubercles, thereby allowing the degree of femoral or acetabular anteversion to be measured along with other important anatomic characteristics, such as the femoral caput collum diaphysis angle, the center edge angle, and the femoral length. MDCT allows the area to be scanned in one volume and patients can be scanned in any position. 3-D reconstructions also may be helpful for evaluation of complex skeletal diseases and for assessing postoperative alignment [22].

MDCT is useful in assessing postoperative anatomy in patients who undergo treatment for developmental dysplasia of the hip (DDH), slipped capital femoral epiphysis (SCFE), and femoroacetabular impingement (FAI) [Fig. 2]. The multiplanar formatting provides an effective assessment of alignment of the proximal femur and acetabulum and optimal assessment of the osseous anatomy.

The labrum, articular cartilage, and intra-articular contents also may be assessed with CT arthrography. The CT arthrogram also can be performed in conjunction with MR arthrography, because 1 to 2 μL of iodinated contrast material usually is used to confirm an intra-articular placement during MR arthrography. This amount of iodinated contrast

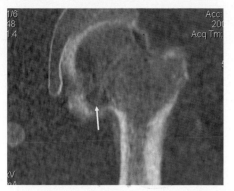

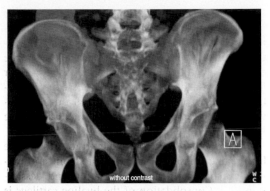

Fig. 2. FAI. Coronal reconstruction after injection of a 3% iodinated contrast mixture shows a bony prominence at the inferior portion of the femoral head-neck junction (*arrow*). This bony prominence was not seen at the time of the initial surgery (osteoplasty of the anterior and superior head neck junction performed because of FAI).

Fig. 3. 3-D volume CT scan of the pelvis shows optimal demonstration of the osseous anatomic relationships.

material provides an appropriate dilute contrast solution that may be examined by CT immediately after the injection of the contrast material. The MDCT examination of the hip requires only a few minutes from the time patients enter the CT suite until the end of an examination, thereby making it possible to scan patients quickly before obtaining the MR imaging examination. When using a protocol that scans through to the knees [Table 2], various anatomic characteristics may be measured (ie, femoral and acetabular version) and calculated (ie, McKibbin index) [23].

In addition to its high-resolution, multiplanar capabilities, MDCT also may be used to map the anatomy three dimensionally. This mapping may be used in THA and hip arthroscopy. Surgical goals generally involve conserving bone stock, optimizing the implant position and fit, appropriately sizing osteochondroplasties, equalization of leg length, creating normal biomechanical orientation, and minimizing complications. CT is useful especially when the hip is prominently dysplastic or when the combined femoral neck and external rotation is greater than 15° [24]. 3-D presentations of anatomy, as can be produced by multislice CT, are effective especially at demonstrating the intricacies of the anatomy [Fig. 3]. MDCT not only is effective for guiding surgical therapy, but it can be useful for postoperative assessment of the osseous anatomy.

Postoperative imaging evaluation

Acetabular labral repair

The acetabular labrum is a fibrocartilaginous rim around the acetabulum that is triangular in cross-section and extends around the periphery of the

Table 2: Protocol for CT scanning of the hip

Parameter	Bone algorithm	Soft tissue algorithm
Kilovoltage	140	140
Milliampere seconds	250	250
Rotation time (s)	0.8	0.8
Collimation (mm)	1.3	2.5
Section thickness (mm)	1.3	2.5–3.0
Reconstruction interval (mm)	0.625	2.5–3.0
Intravenous contrast material	Dilute iodinated contrast (1–2:20) and dilute gadolinium injection (concentration 1:200)	Dilute iodinated contrast (1–2:20) and dilute gadolinium injection (concentration 1:200)

Special Instructions:

    Series #1: scan hips and surrounding soft tissues from superior iliac crests to tibial tubercles (FOV, 36 cm).

    Series #2: small FOV over hip of interest (FOV, 20 cm).

    Series #3 & 4: sagittal and coronal reconstructions.

    Series #5: performed at the workstation or the console: oblique axial reconstruction along the axis of the femoral neck.

    Series #6: performed at the workstation or the console: oblique sagittal plane oriented perpendicular to the femoral neck.

    Note: orient series 5 and 6 parallel and perpendicular to the femoral neck, respectively.

acetabulum, with the exception of the acetabular notch (where it merges with the transverse acetabular ligament) [Fig. 4]. The labrum is bulkier superiorly and posteriorly and thinner at the opposite corner [25,26]. The fibrocartilage of the acetabular labrum is attached directly to the osseous acebtabulum and is composed of heterogeneous fibrocartilage (relative to the fibrocartilage meniscus of the knee) [15].

The acetabular labrum is believed to preserve the hydrostatic pressure within the hip joint and prevent articular cartilage consolidation and dessication [27]. Consolidation of the hyaline cartilage is defined by the compression of cartilage layers with expression of interstitial fluid from the proteoglycan-rich matrix of tissue. The acetabular labrum also contributes to maintaining stability of the femur relative to the acetabulum [27]. The joint capsule inserts onto the acetabular rim at the base of the labrum. At the junction between the labrum and the capsule, there is a perilabral recess [26].

Imaging of the acetabular labrum may be accomplished best with MR arthrography [28]. Knowledge of the arthroscopic classification of acetabular tears also helps match the arthroscopically described abnormality to the MR equivalent appearance [Fig. 5]. With MR arthrography, labral tears are diagnosed when the injected fluid extends into the substance of the acetabular labrum through a surfacing signal abnormality [Fig. 6]. This is applicable regardless of whether or not patient status is preoperative or postoperative. Fluid also may extend between the attachment of the labrum to the transverse acetabular ligament or undermine the labral attachment to the bony acetabulum [Fig. 7]. One of the pitfalls when evaluating the acetabular labrum (postoperative or

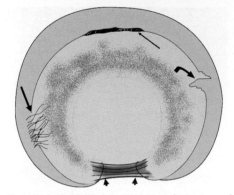

Fig. 5. Acetabular labral tears. An en face view of the acetabulum shows the transverse ligament (*arrowheads*) joining the lunate shaped articular surface of the acetabulum (lighter gray shaded structure) and the two edges of the acetabular labrum (darker gray structure). Some of the various types of acetabular labral tears are illustrated on this image including the peripheral or longitudinal type tear (*small arrow*), the radial fibrillated tear (*large arrow*), and the radial flap tear (*curved arrow*). (*Adapted from* Lage LA, Patel JV, Villar RN. The acetabular labral tear: an arthroscopic classification. Arthroscopy 1996;12:269–72.)

otherwise) is the posteroinferior labral sulcus [Fig. 8] [29]. This sulcus or groove is common and can be mistaken for a labral avulsion if not recognized as a normal variant. Other regions of separation between the labrum and the acetabulum should be interpreted as tears, as most investigators recognize the posteroinferior labral sulcus as the only normal variant of its type [16,30].

Repair of the acetabular labrum may involve partial resection of the labrum, débridement of the torn portion of the labrum, or suture repair with

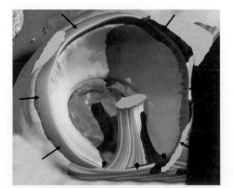

Fig. 4. Diagram of an en face view of the acetabulum and acetabular labrum. The labrum extends around the periphery of the bony acetabulum (*arrows*) with the exception of the acetabular notch where it merges with transverse acetabular ligament (*arrowheads*). (*Adapted from* Primal Pictures Ltd., London, UK; with permission. © Primal Pictures Ltd.)

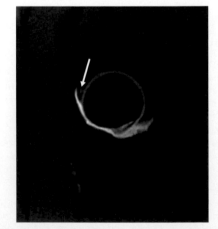

Fig. 6. Labral tear. Sagittal T1-weighted fat-suppressed MR arthrogram shows intra-articular gadolinium and a region of increased signal in the anterosuperior acetabular labrum, consistent with an acetabular labral tear (*arrow*).

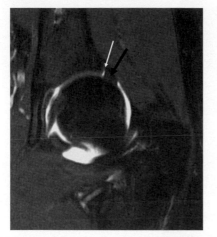

Fig. 7. Labral avulsion. Coronal proton density weighted MR image of the left hip demonstrates fluid (*white arrow*) undermining the superior acetabular labrum (*black arrow*). Arthroscopic evaluation of the left hip showed avulsion of the acetabular labrum from the superior acetabular rim.

marginal convergence of the torn components of the labrum. The majority of labral tears are treated by resection or débridement and although the short-term results are favorable, the long-term re sults are not yet known [31]. Given that the most common method of repairing the acetabular labrum is to remove a portion of it, the normal postoperative MR appearance after partial resection or débridement of the labrum must be considered. The labrum may appear truncated or partially or completely absent after surgical treatment for an acetabular tear [Fig. 9]. Fluid that either undercuts the labrum or extends into its substance, however, is indicative of a persistent tear or a repeat injury to the labrum.

Inflammatory changes in and around the perilabral recess manifest as an anatomic adherence between the perilabral joint capsule and the labrum itself [Fig. 10A]. Normally, this recess is present and is defined on MR arthrography with fluid extending between the acetabular labrum and the joint capsule [see Fig. 10B]. The adhesion of the joint capsule to the labrum is common in the superolateral perilabral recess and the capsule may appear scarred and adherent to the labrum on MR imaging and arthroscopy. Adhesion of the capsule to the labrum may have an effect on the ability of the labrum to maintain its hydrostatic seal and thereby could expose the articular cartilage to increased stress. In patients who have perilabral adhesions that are examined by arthroscopy, the joint capsule can be seen as adherent to the labrum. This is associated commonly with erythema and limited capsule motion across the labrum in the region of adhesions.

Focal separation of the acetabular labrum from the bony acetabulum rarely may be seen after a hip arthroscopy procedure. This can be seen on follow-up MR imaging and with follow-up arthroscopy, as a focal perforation in the base of the acetabular labrum may result from the placement of the portal through the substance of the labrum. Although other complications of arthroscopy, such as injury to the lateral femoral cutaneous nerve or chondral injuries, are more common, radiologists should be aware of this iatrogenic cause of labral perforation.

Normal variant appearances of the acetabular labrum also should be taken into consideration during postoperative imaging of the hip. The acetabular labrum has a triangular appearance in most people. This typically transitions to a rounded appearance then to an irregular appearance and may appear absent in a certain portion of older patients [32]. Signal alteration within the acetabular labrum also may be seen. Signal changes typically start in the anterior and superior portion of the acetabular labrum and progress to other parts of the labrum as patients age [32]. Variation in the typical appearance of the acetabular labrum relates partially to patient age and to traumatic pathology, and this should be taken into consideration when interpreting imaging studies of patients who have hip discomfort.

Femoral head-neck junction osteoplasties

Open surgical technique

Proximal femoral osteoplasty is performed to correct osseous abnormalities that cause impingement of structures in or around the hip joint. There are open and arthroscopic techniques for performing these osteoplasties [14]. Open procedures involve

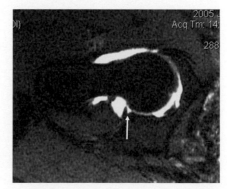

Fig. 8. Sublabral sulcus. Oblique axial T1-weighted fat suppressed image obtained after injection of intra-articular gadolinium contrast shows a sublabral groove or sulcus (*arrow*) at the posteroinferior portion of the acetabular labrum.

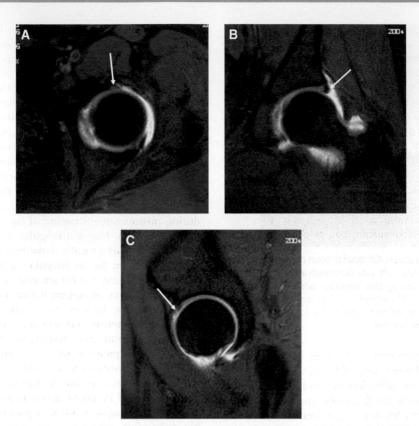

Fig. 9. Débrided labral tear. Axial (*A*), coronal (*B*), and sagittal (*C*) T1-weighted fat saturated MR images obtained after the direct administration of a dilute (1:200) gadolinium solution shows truncation and irregularity of the anterosuperior portion of the acetabular labrum (*arrows*). This is consistent with prior arthroscopic débridement of a labral tear in this region.

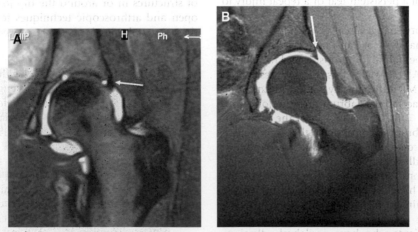

Fig. 10. Superolateral hip joint capsule. (*A*) Coronal T2-weighted fat suppressed image obtained after the injection of 10 cc of dilute gadolinium into the hip joint shows that the superolateral hip joint capsule is adherent to the lateral portion of the superior acetabular labrum (*arrow*). (*B*) Coronal T1-weighted fat suppressed image obtained after the injection of 10 cc of dilute gadolinium into the hip joint shows a normal perilabral recess with fluid extending between the medial border of the superior hip joint capsule and the lateral portion of the superior acetabular labrum (*arrow*).

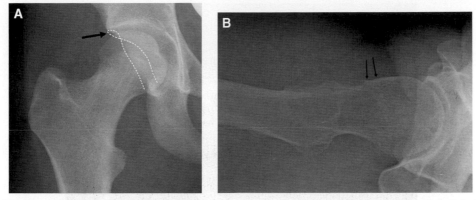

Fig. 11. "Cross-over" sign of FAI. (*A*) Preoperative anteroposterior radiograph of a patient who had clinical signs of FAI syndrome shows a cross-over of the superior aspect of the posterior acetabular wall (*arrow*). The acetabular margin is outlined (*white dashed line*), delineating normal mid and inferior acetabular relationships. The superior aspect of the acetabulum reverses its orientation, however, such that the posterior acetabulum lies medial to the anterior wall. (*B*) Preoperative lateral radiograph demonstrates abnormal prominence of the anterior femoral head neck junction (*arrows*).

placing patients in the lateral decubitus position with the hip exposed anteriorly and dislocated in the same direction. Care is taken not to disrupt the external rotator muscles and the vascular supply (medial circumflex femoral artery) that parallel these muscles. This technique also allows a complete view of the femoral head and the acetabulum. After the hip is dislocated, the femoral head-neck junction may be inspected for prominent regions and decreased head-neck cutback zones, and cartilage surfaces are inspected for damage. The treatment for osseous impingement is to remove the nonspherical portions of the femoral head and neck via an open surgical osteochondroplasty [Figs. 11 and 12]. Acetabular sources of impingement, such as acetabular retroversion, prominence of the lateral acetabular rim, and acetabular protrusio, also may be treated either by osteochondroplasty of the lateral acetabular rim or acetabular osteotomy. In many cases, it is necessary to address both sites of impingement [see Figs. 11 and 12] [14]. After surgical resection, it is important to observe the full range of motion of the hip to ensure that there are no residual regions of impingement.

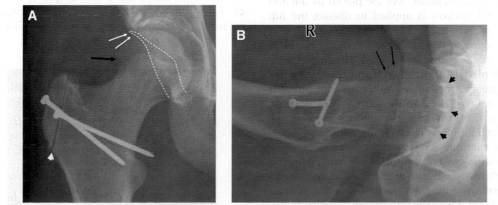

Fig. 12. Status post osteoplasties for FAI. (*A*) Anteroposterior conventional radiograph of the patient shown in Fig. 11 is obtained postoperatively after undergoing a surgical hip dislocation with osteoplasties of the femoral head-neck junction (*black arrow*) and superolateral anterior acetabular rim (*white arrows*) and a greater trochanteric osteotomy (*white arrowhead*) with screw fixation (done as part of the surgical dislocation procedure). The acetabular rim again is demonstrated by the white dashed line. Note the lack of cross-over along the superior margin of the osseous acetabulum after the prominent anterosuperior acetabular rim has been resected. (*B*) Lateral postoperative radiograph shows decreased prominence of the anterior femoral head neck junction because of the prior osteoplasty (*arrows*). The acetabular rim osteoplasty also has been performed and the labrum tacked down (*arrowheads*).

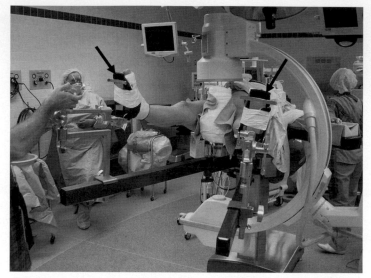

Fig. 13. Patient positioning for hip arthroscopy. Patient is prepared for hip arthroscopy and positioned in a modified supine position on a common fracture table. The traction boots are applied (*arrows*) with careful attention to ensure sufficient padding is present and that the patient is protected adequately against abrasion or mechanical injuries.

Arthroscopic technique

Hip arthroscopy may be used for resection of prominent portions of the femoral head-neck junction and is less invasive than hip arthrotomy. Although effective in many circumstances, hip arthroscopy has contraindications, including patient obesity, the presence of open wounds or cellulitis, arthrofibrosis, and very advanced disease states. The arthroscopic procedure is performed in patients who are in a modified supine position on a fracture table. Patients' feet are placed in traction boots and traction is applied to distract the hip joint [**Fig. 13**]. This usually results in 8 to 10 mm of distraction and enables the internal portions of the joint to be examined [**Fig. 14**]. The anterolateral portal is placed 1.0 cm anterior and 1.0 cm superior to the greater trochanter and the anterior portal is placed 2.0 cm lateral to the anterior superior iliac spine at the same superior/inferior level as the anterolateral portal. Through the arthroscopic portals, the anatomy is inspected [**Fig. 15**] and regions of abnormal osseous prominence are identified. The arthroscopic bur is used to resect the osteocartilaginous prominence associated with patients' FAI [**Fig. 16**]. Once the arthroscopic osteochondroplasty is performed, the range of motion at the hip joint is examined to ensure that the osseous prominences have been resected adequately [**Fig. 17**]. If residual osseous impingement is identified by arthroscopic examination during range of motion movements, additional osteoplasty or osteochondroplasty may be necessary to remove the regions of impingement. Evidence of the head-neck junc-tion osteoplasty may be seen with postoperative conventional radiography and with cross-sectional imaging [**Fig. 18**].

Many patients presenting with osseous impingement also have other pathology, such as acetabular labral tears and flexion contractures at the hip joint. The labral tears are repaired or débrided in conjunction with the osteoplasty, and an iliopsoas tendon release also can be performed in patients who have flexion contractures of the hip or signs of femoral nerve entrapment. The imaging appearance of the iliopsoas release should be recognized to avoid misinterpretation. Acutely, the iliopsoas

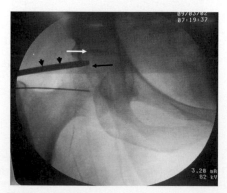

Fig. 14. Hip joint distraction. Frontal fluoroscopic image shows the distraction at the right hip joint as represented by the distance between the superior portion of the femoral head (*black arrow*) to the inferior border of the acetabulum (*white arrow*). The anterolateral arthroscope is in place (*black arrowheads*).

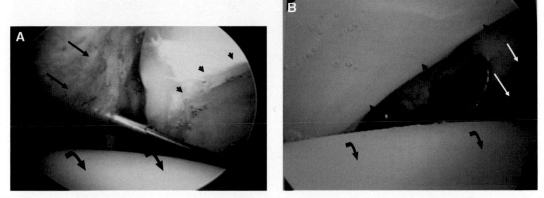

Fig. 15. Hip arthroscopy. (*A*) Arthroscopic view of the hip from the anterolateral portal (looking anteriorly) identifies the iliofemoral ligament as the anterior border of the joint (*arrows*) and the acetabular labrum (*arrowheads*). The hyaline cartilage of the articular surface of the femoral head also is seen at the inferior margin of the FOV of the scope (*curved arrows*). (*B*) Arthroscopic view of the hip from the anterior portal (looking posteriorly) is performed initially to check the position of the anterolateral portal. The ischiofemoral ligament is seen adjacent to the portal (*white arrows*) and the acetabular labrum again is noted at the margin of the posterior acetabulum (*black arrowheads*). The articular surface of the femoral head again is noted (*curved black arrows*).

release can result in edema and fluid around the distal portion of the tendon. The fluid commonly tracks from the point of release (adjacent to the joint) superiorly and may surround the entire distal tendon [**Fig. 19**]. The fluid also may track superiorly to be located between the medial iliopsoas tendon and the external iliac artery and vein [see **Fig. 19**]. The chronic appearance of an iliopsoas release is characterized by an absence of fluid and the presence of iliopsoas muscle atrophy [**Fig. 20**]. This is a normal postoperative appearance and should not be confused with disuse atrophy, denervation, or postmyositis atrophy.

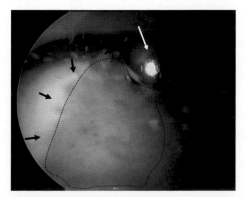

Fig. 16. Osteoplasty for FAI. Arthroscopic image taken during bur resection of a prominent femoral head-neck junction demonstrates the bur (*white arrow*) used to remove the osseous prominence under direct visualization of the camera. The edge of the articular surface can bee seen (*black arrows*) adjacent to the medullary bone that has been exposed by the osteoplasty (area within the dotted line).

Hip fracture repair

The two primary elements that should be present for adequate fracture healing to occur are adequate reduction and internal fixation [33]. Inadequate reduction increases the risk of avascular necrosis (AVN) and unstable fixation [34]. After fracture fixation, the intraoperative fluoroscopic anteroposterior view should demonstrate a caput collum diaphysis angle ranging between 130° and 150° [**Fig. 21**] [35]. This angulation purposefully is valgus to reduce the risk of AVN and maximize the stability of the fixation. On the lateral view, the alignment of the femoral head to the shaft should be linear [see **Fig. 21**] [35]. On the lateral view, the screw should be placed in the center of the femoral neck and through the center (or slightly posterior) portion of the femoral head [36]. A 1995 study by Baumgaertner and coworkers support a distance from the tip of the dynamic hip screw to the femoral head apex of 25 mm or less [37]. The femoral head apex is determined by the junction of the subchondral femoral head and a line drawn parallel along the center of the femoral neck. Care must be taken to scrutinize the position of the distal tip of the screw on the lateral view adequately as the proximal femur can be difficult to visualize adequately and the interobserver agreement as to the determination of this position is reported as poor [9].

The goal of a hip fracture repair is to place the proximal femur and the femoral fixation construct in such a position that the repair is stable and the fracture heals without resultant deformity. The complications of a femoral neck fracture in children are similar to complications seen in other

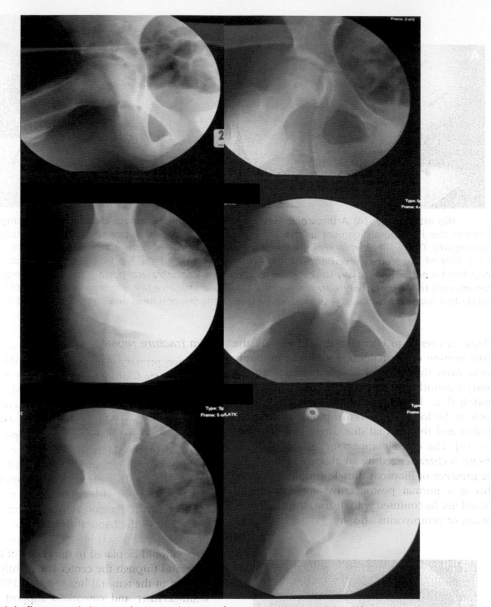

Fig. 17. Multiple fluoroscopic images show good range of motion and no evidence of osseous impingement. The range of motion must be examined intraoperatively to ensure there is no residual impingement.

types of proximal femoral stabilization, such as DDH repair and pinning for SCFE [38]. Postsurgical deformities may occur after fixation and additional manipulation may be necessary to treat these regions of altered anatomy. In addition to the deformities that may cause biomechanical compromise resulting from abnormal femoral or acetabular version, other processes, such as leg length discrepancies, abnormal varus, or valgus angulation, and disruption of the proximal femoral epiphysis may give rise to proximal femoral deformities [Figs. 22 and 23] [39]. The post-traumatic alteration in anatomy can give rise to early-onset de-

generative arthritis. Additionally, acetabular labral tears rarely occur in the absence of bony abnormalities and this combination of pathology can contribute to a painful post-traumatic or postsurgical hip [40].

Core decompression

AVN is a common condition affecting between 10,000 and 20,000 individuals in the United States annually [41]. Various risk factors for AVN are identified, including trauma, steroid medications, systemic lupus erythematosis, and pancreatitis. The primary method of treating AVN is early identifica-

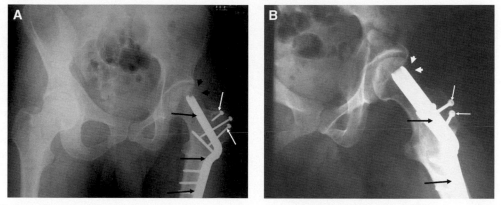

Fig. 18. Osteochondroplasty for FAI. Anteroposterior (*A*) and frog-leg lateral (*B*) views of the left hip show a blade plate and screw construct within the left proximal femur (*black arrows*), greater trochanteric screw fixation (*white arrows*), and postoperative evidence of the femoral head neck junction osteochondroplasty (*white arrowheads*).

tion by MR imaging and CD of the femoral head. The effect of CD is to decrease the intraosseous pressure and to promote the revascularization of the femoral head. Overall, there is controversy about the role of CD in the treatment of early AVN [42]. Compared with conservative management, however, there are some reports that CD produces better treatment results in early stage AVN [43].

There are different techniques that are used for CD, all of which involve accessing the femoral head to create a void within the cancellous bone [Fig. 24]. Accessing the femoral head with a Kirschner wire, hollow trephine, or screw is used and some procedures are performed in conjunction with bone grafting into the void. These techniques are used with varying results but, in combination

with early detection with MR imaging, the results of CD generally are improved.

The postoperative status of the CD with or without the grafting procedure can be monitored and the degree of involvement of the articular surface of the femur can be measured. The degree of involvement of the articular surface is used as a preoperative predictor for subchondral collapse. In 1990, Beltran and colleagues divided the degree of articular involvement of the femoral head into four categories: (A) no AVNs, (B) less than 25% involvement of weight bearing portion of the femoral head, (C) involvement of 25% to 50%, and (D) involvement of more than 50%. The rate of collapse varied from no collapse in categories A and B to collapse seen in 43% of the hips in category C to 87% of the hips in category D [44].

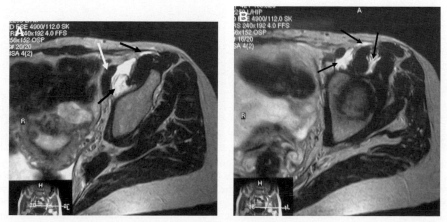

Fig. 19. Iliopsoas tendon release. (*A*) Axial fast spin-echo T2-weighted image just superior to the left hip joint in a patient who recently has had an iliopsoas tendon release shows fluid surrounding the anterior and medial portions of the iliopsoas tendon (*black arrows*). The fluid also lies immediately adjacent to and lateral to the external iliac neurovascular bundle (*white arrow*). (*B*) Axial fast spin-echo T2-weighted image just inferior to the image shown in (*A*) shows fluid surrounding the anterior, medial and lateral portions of the iliopsoas tendon (*black arrows*).

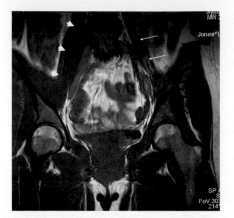

Fig. 20. Iliopsoas atrophy after tendon release. Coronal T1-weighted MR image of the pelvis shows atrophy of the left psoas major muscle (*white arrows*) compared with the normal right psoas major (*white arrowheads*). This patient had undergone an iliopsoas tendon release months before this examination.

The amount of edema associated with the double line sign seen in AVN also may be used as a predictor for the chance of success in CD of the femoral head [45]. In a 2004 study, Radke and coworkers find that the presence of edema associated with the double line sign of AVN is a predictor for progression to THA. The findings of extent of involvement of the articular surface and the presence of edema along with the double line sign may or may not have similar prognostic usefulness in postoperative hips. This needs to be examined by additional studies that strive to re-

late the postoperative appearance after CD to the time course and degree of disease progression. The deformity of the femoral head along with the presence of subchondral collapse, joint space narrowing, osteophyte formation, and the progression of degenerative arthritis may be visualized clearly on follow-up radiographic or cross-sectional imaging. These various modalities may be used as a method of monitoring patients and, in conjunction with the clinical examination, contribute substantially in determining whether or not a patient's treatment regimen should progress to surgical replacement of the involved joint.

Total hip arthroplasty

As discussed previously, the number of THA procedures has increased to the point where this is a routine treatment for patients who have severe hip joint arthritis or in patients who have sustained an irreparable hip fracture. The complications of THA generally are well recognized and include loosening of the prosthesis, periprosthetic fracture, heterotopic bone formation, postoperative infection, trochanteric bursitis, and osteolysis from foreign body granulomatosis. These complications may vary in severity and often necessitate revision arthroplasty.

Imaging of total hip arthroplasty

The imaging evaluation of symptomatic patients who have undergone THA includes primarily con-

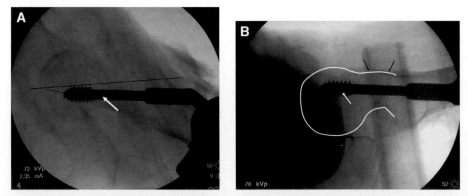

Fig. 21. Dynamic hip screw. (*A*) Anteroposterior fluoroscopic image taken in the operating room during a proximal femoral fracture fixation shows the dynamic hip screw (*white arrow*) located within the proximal femur. The proximal femur was fixed optimally in slight valgus, as the caput collum diaphysis (neck shaft) angle was 135°. The distance from the tip of the screw to the femoral apex (*dotted black line*) measured 18 mm. The femoral apex is designated by the subchondral point of the femoral head as it intersects a line drawn down the center of the femoral neck (*solid black line*). (*B*) Lateral fluoroscopic view of the proximal femur taken after placement of a dynamic hip screw (*white arrow*) shows the position of the screw relative to the femoral head and neck (outlined by solid white line). Ideally, the dynamic hip screw should be placed along the longitudinal axial axis of the femoral neck and placed in the center of the femoral head or slightly posterior to center. The dynamic hip screw in this case is angled slightly anterior of the optimal position and the femoral neck is convex anteriorly (*black arrows*), which can result in femoral retrotorsion.

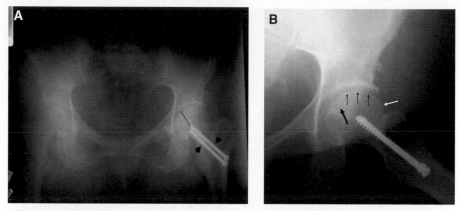

Fig. 22. Pin fixation of femoral neck fracture. (*A*) Anteroposterior radiograph of the pelvis shows two fixation screws in the left femoral neck (*black arrowheads*) that are located toward the inferior portion of the femoral neck and more than 25 mm away from the femoral head apex (*dotted black line*). This position is because of placement of the pins in childhood when the patient sustained a femoral neck fracture. The caput collum diaphysis angle is normal and there is no evidence of prior AVN. (*B*) Frog-leg lateral radiograph of the left hip again shows the fixation pins (*black arrowhead*). There also is evidence of femoral head flattening superiorly (*small black arrows*) but with femoral head sclerosis. A subchondral cyst (*large black arrow*) is noted in the medial portion of the subchondral femoral head along with very early osteophyte formation (*white arrow*).

ventional radiography, nuclear medicine studies with technetium and gallium scanning, joint aspiration, and arthrography [46–49]. Traditionally, cross-sectional imaging was not used because of imaging artifacts caused by the metal prostheses (beam hardening and streak artifacts with CT and metal susceptibility artifacts with MR imaging). Various strategies have been used to decrease the MR imaging artifacts caused by a variety of metal orthopedic devices [10,50–53]. These are discussed in more detail later.

MT imaging also may be useful in the imaging evaluation for loosening of hip arthroplasties. Lower field-strength MR imaging units (ie, less than 1.0 Tesla) may be the most advantageous in this evaluation, because less susceptibility artifact results from lesser gradient strengths [54]. Low-field MR imaging is shown to be able to depict loosening around the femoral component of a THA, but the visualization around the acetabular component is less optimal [54,55]. Higher magnetic field strengths often are associated with proportionately more prominent artifacts [Fig. 25]. This contributes to the common belief that MR imaging is not appropriate for detecting periprosthetic pathology, such as loosening or particle inclusion disease in patients who have arthroplasties, especially THAs. Various MR imaging techniques, however, have been developed to combat the susceptibility artifacts seen with MR imaging.

MR imaging is shown to be accurate in detecting and sizing areas of osteolysis [56] and in evaluating the surrounding soft tissue structures [Fig. 26] [6].

As discussed, MR imaging after arthroplasty poses particular challenges because of the prominent artifact from the metal prostheses. The different types of metal used in hip arthroplasty demonstrate variations in the severity of artifact—stainless steel causes severe distortion and titanium or oxidized zirconium causes less severe distortion [57].

Because the artifact from metal partly is the result of incorrect spatial encoding resulting from frequency misregistration, widening the receiver bandwidth decreases the apparent size of artifact. This occurs because the widened bandwidth allows a broader range of frequencies to be sampled, so that the difference between any two frequencies is a smaller percentage of the total range. In one study, widening the receiver bandwidth from 16 kHz to 64 kHz reduces metal artifact by 60% [58]. In addition to sampling over a broader range of frequencies, widening the receiver bandwidth also serves to decrease interecho spacing. A shorter echo time allows less time for intravoxel dephasing than a long echo time, and is, therefore, less sensitive to metal artifact. The downside of widening the receiver bandwidth is that it causes a decrease in SNR [59]. To compensate for the loss of SNR, increasing the number of excitations (NEX) may be necessary.

The echo train length (ETL) also is shown to be a factor in the severity of artifacts [59]. Increasing the number of refocusing pulses (increasing ETL) decreases the time for intravoxel dephasing to occur.

The misregistration from metal implants occurs between slices (in the z-plane) and in the frequency encoding direction (y-plane). Similar to the receiver

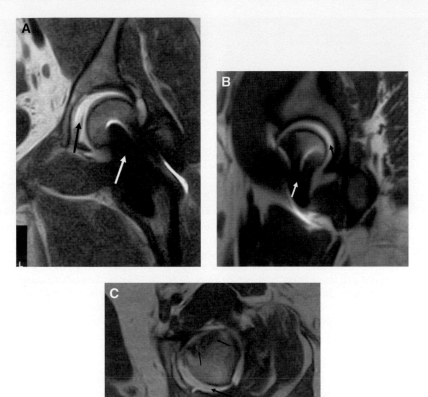

Fig. 23. Nonspheric femoral head secondary to childhood fracture fixation. (*A*) Coronal T1-weighted MR image from a left hip MR arthrogram shows the metallic artifact from the fixation screw (*white arrow*) along with widening of the medial hip joint space (*black arrow*). (*B*) Sagittal T1-weighted MR image again demonstrates the metallic artifact (*white arrow*) but shows that the joint space is widened posteriorly (*black arrow*). (*C*) Axial T1-weighted MR image shows the joint space widening posteriorly (*large black arrow*), femoral head flattening, and degenerative change anteriorly (*small black arrows*). This axial image also provides an optimal view of the nonspherical femoral head that resulted from the childhood fracture fixation. This nonsphericity produced an asymmetric position of the femoral head within the acetabulum that resulted in atypical degenerative changes. These findings were corroborated by a hip arthroscopic examination. An anterosuperior labral tear that was identified at MR imaging and arthroscopy is not shown.

bandwidth, the slice-select bandwidth can be widened by increasing the magnitude of the gradient and decreasing the time in which the gradient is applied. For example, a 20% increase in bandwidth can be achieved by multiplying the gradient magnitude by a factor of 1.2 and decreasing the time the gradient is applied by a factor of 1.2 [58]. This decreases misregistration between slices.

An additional technique used by some investigators to reduce metal artifact is called view angle tilting (VAT) [58,60,61]. This is achieved by re-applying the slice-select gradient at time of readout, so that the slice-select gradient and the frequency-encoding gradient are applied simultaneously. VAT causes the spins to be aligned essentially tangential to the frequency plane at time of readout and is shown to reduce susceptibility artifact [56,60,61]. Image blurring is a problem resulting from this technique, and different methods are attempted to reduce blurring [56,61,64]. In one study, a combination of widening the receiver bandwidth, widening the slice-select bandwidth, and applying VAT resulted in decreased artifact compared with either widening the receiver bandwidth or VAT alone [58].

Overall, with use of a few modifications, MR imaging can be useful and accurate in evaluating the periprosthetic osseous and soft tissue structures

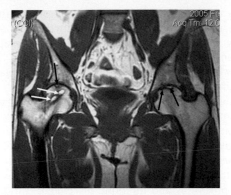

Fig. 24. Core decompression (CD). Coronal T1-weighted MR image of the pelvis shows a linear signal alteration in the right femoral head and neck (*white arrows*), consistent with the patient's prior CD. The high signal within the region of CD represents marrow fat that has grown back into the region of the CD. There also are serpiginous areas in the subchondral regions of the femoral heads (*black arrows*), consistent with AVN.

after hip arthroplasty. In the future, MR imaging may become the image modality of choice for the painful hip prosthesis.

Common pathology in total hip arthroplasty

Soft tissue abnormalities

Although the imaging work-up usually is directed toward detecting hardware failure, various soft tissue abnormalities may be the actual cause of hip pain. MR imaging may be used to assess processes that occur around a metal implant, including adductor tendon rupture and trochanteric bursitis [62]. These soft tissue abnormalities may be the cause of pain, especially if a transgluteal approach is used [63]. MR angiography also is used to detect deep venous thrombosis in patients who have undergone THA [51].

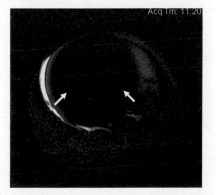

Fig. 25. Metallic artifact. Axial T2-weighted MR image of the thigh demonstrates a prominent metallic artifact (*white arrows*) from a long stem hip prosthesis.

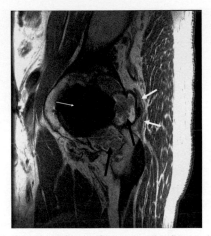

Fig. 26. Particle inclusion disease. Sagittal T1-weighted image in a patient who had a THA shows the loss of signal from the metallic artifact caused by the prosthetic femoral head (*small white arrow*) along with the mixed signal of the particle inclusion disease (*black arrows*). The superior most component of the particle inclusion disease is minimally displacing the sciatic nerve (*large white arrows*) posteriorly. (Courtesy of Hollis G. Potter, MD, Hospital for Special Surgery, New York, NY.)

Loosening

Loosening of a metallic hip prosthesis is the most common indication for revision arthroplasty and typically is determined based on the radiographic appearance of the prosthesis relative to the surrounding bone or polymethyl methacrylate. A loose prosthesis is characterized either by a progressively increasing radiolucency at the bone-cement or prosthesis-cement/bone interface or by a change in the position of the prosthesis. A subtle increase in the radiolucent interface zone, however, is not necessarily a sign of loosening [64].

When patients present with a painful prosthetic hip, the prospect of loosening should be suspected immediately, but before this diagnosis can be made, it should be corroborated by the appropriate imaging examinations. In addition to conventional radiography, other modalities may be used, including arthrography, bone scintigraphy, and MR imaging [65,66].

On conventional radiographs, a lucent zone greater than 2 mm around the prosthesis or at the cement/bone interface is a common manifestation of loosening [Fig. 27] and lucent zones of less than 2 mm cannot be dismissed completely as normal postoperative changes. Caution must be used when invoking the possibility of loosening and the most important diagnostic tool in an armamentarium is the prior comparison radiographs. Some lucent zones seem too wide but do not change over time. Revision arthroplasties also may have a wider

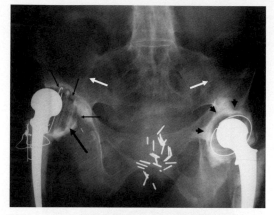

Fig. 27. Loosened acetabular component. Anteroposterior conventional radiograph of the pelvis shows bilateral THAs with cemented prostheses. Note lucency (*small black arrows*) surrounding a displaced and rotated right acetabular cup (*large black arrow*). By comparison, note the normal left acetabular cup (*black arrowheads*). Incidentally noted is sacroiliac joint fusion (*white arrows*) and confluent ossification of the spine in this patient who has ankylosing spondylitis.

radiolucent zone than primary arthroplasties. The possibility of loosening in the absence of clinical symptoms must be considered with skepticism. Other signs of loosening on conventional radiographs include fracture of a prosthesis itself, fracture of the cement around a prosthesis, and the development of osseous sclerosis adjacent to the distal tip of a prosthesis. Occasionally, a prosthesis

may move or have various degrees of rotation from one radiograph to another. Prosthetic wear also may be apparent on conventional radiography when there is particle inclusion disease [Fig. 28] or when there is increasing number of metallic fragments in and around the joint.

The imaging investigation for loosening also includes arthrography. This may be performed in isolation or in conjunction with joint aspiration or synovial biopsy when the possibility of infection is present. A loose prosthesis typically demonstrates contrast below the intertrochanteric line between the prosthesis and the bone or cement. Although arthrography previously has been shown to be accurate in the determination of prosthetic loosening, false negative examinations may occur when inflammatory debris prevents contrast material from entering the interface between the prosthesis and the adjacent bone or cement.

Infection

Along with loosening, infection is one of the most common complications in patients who have hip arthroplasties. Imaging findings in patients who have a low-grade infection may mimic loosening. When the infection is more severe, it often is associated with a hip effusion, prominent bone destruction or sclerosis, and pain. Infection also may mimic aggressive granulomatosis (particle inclusion disease), because osseous resorption may

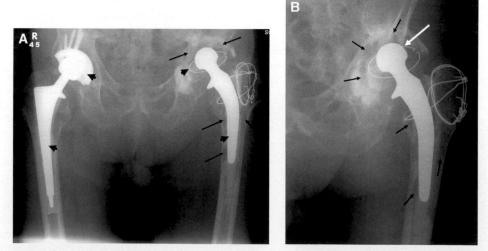

Fig. 28. Loosened acetabular component. (A) Anteroposterior radiograph demonstrates bilateral THAs (*arrowheads*). The right hip prosthesis is normal in appearance, but there are multiple lucencies surrounding the acetabular and femoral components (*arrows*). There also is loosening and lateral subluxation of the left acetabular component. (B) Anteroposterior radiograph of the left hip taken 6 months later shows progression of the lucencies surrounding the femoral component (*black arrows*). There also is progressive lateral subluxation of the acetabular component, so that the lateral portion of the articulated femoral head is aligned with the lateral ilium (*white arrow*).

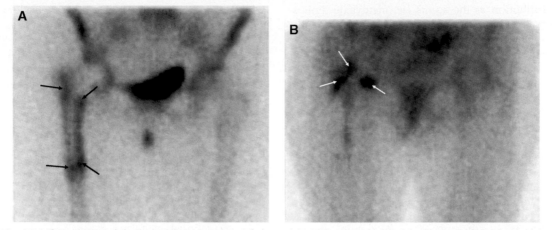

Fig. 29. Infected THA. (A) Anteroposterior view of the pelvis and proximal femora taken after a 3-hour delay during a bone scan performed with technetium Tc 99m methylene diphosphonate shows increased radiotracer uptake around the femoral component of the THA (*black arrows*). (B) Anteroposterior view of the pelvis and proximal femora in the same patient as demonstrated in (A) taken during a gallium citrate Ga 67 citrate scan shows radiotracer accumulation around the region of the right hip joint (*white arrows*). The presence of the radiotracer within the joint is indicative of infection. (Courtesy of Christopher J. Palestro, MD, Long Island Jewish Medical Center, New Hyde Park, NY).

occur around the prosthesis and can simulate the scalloped appearance of resorption seen in periprosthetic granulomatous involvement.

In the imaging evaluation of a suspected arthroplasty infection, conventional radiography is used and can show findings, such as periprosthetic resorption of bone, loosening of the prosthesis, and osseous changes consistent with osteomyelitis. Cross-sectional imaging can be used to confirm or refute findings seen on radiographs, assess for soft tissue involvement, define the degree of bony destruction better, and evaluate more definitively the osseous structures for evidence of osteomyelitis. Radionuclide bone scans may demonstrate findings in infection that are similar to those found in aseptic loosening of the prosthesis. Correlation with dedicated scans designed to assess for infection may be helpful in differentiating infection and loosening. The combination of technetium and gallium scanning may not always be helpful in differentiating infection from joint arthroplasty loosening, because periprosthetic bone can exhibit uptake with both scenarios. This results from gallium's proclivity to be taken up in bone in a similar

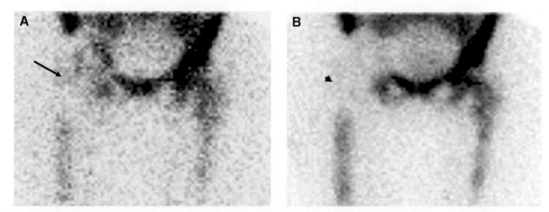

Fig. 30. Infected THA. (A) Anteroposterior view of the pelvis and proximal femora in a patient who had a right THA taken after administration of indium In 111–labeled WBCs shows a faintly increased region of uptake in the region of the right proximal femur (*arrow*). (B) Anteroposterior view of the pelvis and proximal femora in a patient who had a right THA taken after administration of technetium Tc 99m sulfur colloid shows no uptake in the region of the right hip (*arrowhead*). Therefore, the presence of increased radiotracer uptake on the WBC scan and absence of radiotracer uptake on the sulfur colloid scan are consistent with the presence of infection. (Courtesy of Christopher J. Palestro, MD, Long Island Jewish Medical Center, New Hyde Park, NY).

distribution as the agents (ie, methylene diphosphonate) used for bone scanning. Occasionally, gallium scanning may show uptake within the joint itself, which is highly indicative of infection [Fig. 29]. A standard way of analyzing the combination of bone scan and gallium studies includes considering the scenario negative for infection when the gallium scan is normal, when the radiotracers are congruent, or the intensity of the gallium is less than that of the bone scanning agent. The scenario is considered positive for infection when the intensity of gallium exceeds that of the bone agent or when they are spatially incongruent [67].

Radionuclide scanning using sulfur colloid imaging combined with technetium- or indium-labeled white blood cells (WBCs) may be helpful adjuncts to the anatomic imaging techniques for differentiating between loosening and infection Because neutrophils are present with an infection, labeled WBC imaging is the most sensitive technique for detecting neutrophil-mediated inflammatory changes. The radiotracer labeled neutrophils collect in the region of infection but not in the normal marrow to nearly the same degree. The labeled sulfur colloid is present in the normal marrow but not in the regions of infection. The degree of radiotracer uptake along with the type of radiotracer that collects in the specific location of interest provide important clues to the cause of the patient's symptoms or the findings on the anatomic imaging studies [Figs. 30 and 31].

Regardless of the findings on any particular examination, the information derived from an imaging evaluation must be evaluated collectively along with any available clinical information, because some conditions, such as cellulitis or inflammatory arthritis, may mimic the presence of infection [68]. In equivocal cases, fluoroscopic or sonographically guided joint aspiration or synovial biopsy may be used to assess for infection further.

Particle inclusion disease

Focal osteolysis around the prosthesis was described first by Charnley in the 1960s [73], was believed caused by the cement used to stabilize the prosthesis, and originally was referred to as cement disease. This process subsequently has been called particle inclusion disease or histiocytic response. Any small particles, such as cement, metallic fragments, or small particles of polyethylene, can cause this osteolytic response and this process can occur around any arthroplasty. Any process that accelerates the wear of the prosthesis (ie, an eccentric position of the prosthetic femoral head within the acetabular component) can put patients at risk for particle inclusion disease [69]. With an increase in the use of cementless fixation, polyethylene wear debris is now the most common cause of particle inclusion disease.

The histiocytic response incited by the small foreign particles results from the macrophage reaction to these particles. These granulomatous lesions present as radiolucencies that surround the arthroplasty components. This condition typically occurs 1 to 5 years after surgery and is characterized by

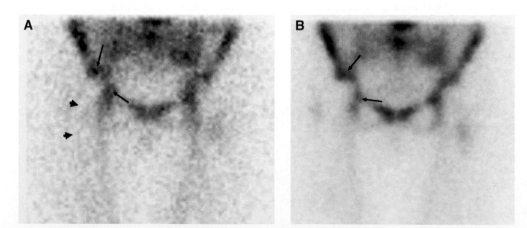

Fig. 31. Noninfected THA. (A) Anteroposterior view of the pelvis and proximal femora obtained during an indium In 111–leukocyte-labeled radionuclide scan shows the presence of a focal region of decreased radiotracer uptake corresponding to the patient's THA (arrowheads). There also is increased radiotracer uptake around the acetabular component of the prosthesis (arrows) that is slightly more than the contralateral side. (B) Anteroposterior view of the pelvis and proximal femora on the same patient during a technetium Tc 99m sulfur colloid scan shows the slightly asymmetric increased uptake around the acetabular component of the right THA (arrows). This uptake indicates the presence of normal marrow in the region and supports the presence of an aseptic process. (Courtesy of Christopher J. Palestro, MD, Long Island Jewish Medical Center, New Hyde Park, NY).

scalloped areas of bony resorption around the prosthesis rather than linear areas around the prosthesis seen with mechanical loosening [see Fig. 28]. Soft tissue masses or fluid representing pseudobursae also may be present around the joint prosthesis.

The bone surrounding the regions of osseous resorption is at increased risk of fracture and the disease process should be followed with serial radiographs or cross-sectional imaging to determine the rate of growth of the granulomatous regions and to assess the degree of fracture risk. Although the degree of involvement is assessed effectively and rapidly with conventional radiography, cross-sectional imaging, such as MR and CT, may be used to monitor progression of the bony resorption [see Fig. 26]. Particle inclusion disease typically is seen on MR imaging as having low intensity signal on the T1-weighted images, intermediate signal intensity on the T2-weighted images, and irregular peripheral enhancement after contrast administration [70]. Periprosthetic fractures also are detected effectively with MR imaging [71].

Summary

The hip often is a diagnostic quandary preoperatively and, relative to other large joints, has received much less attention in terms of postoperative imaging. A complete postoperative evaluation may include the use of many modalities, such as conventional radiography, MDCT, MR imaging, and radionuclide imaging. Many types of surgical procedures are performed on the hip, including acetabular labral repair, osteoplasties, osteotomies, fracture fixation, CD, and THAs. Some of these procedures (ie, fracture fixation and THA) can cause prominent postoperative imaging artifact on certain imaging modalities and, until the advent of MDCT and dedicated MR imaging protocols, cross-sectional imaging of these prostheses was of limited value. The increasing prevalence of open hip surgery and hip arthroscopy has given rise to an increasing variety of procedures and an increased need for postoperative imaging. As with other types of postoperative imaging, knowledge of the type and technical details of the surgical procedure help prescribe which imaging modality demonstrates the findings of interest best. Each type of procedure has certain intricacies associated with it and certain postoperative anatomy that is within the accepted range of normal. Patients' clinical examination and type of surgical procedure that has been performed determine which imaging strategy to use to investigate patient postoperative condition. The ability to direct this workup and to determine what pathology exists, if any, is the task of the individuals directing the imaging evaluation. The

more knowledge these individuals have about the procedures performed and which imaging modalities to use when, the more accurate a work-up is.

References

[1] Byrd JWT. Hip arthroscopy. In: Proceedings of the Arthroscopy Association of North America. 2003. Rosemont (IL): Arthroscopy Association of North America; p. 24–7.

[2] Radke S, Kirschner S, Seipel V, et al. Magnetic resonance imaging criteria of successful core decompression in avascular necrosis of the hip. Skeletal Radiol 2004;33:519–23.

[3] Borrelli Jr J, Ricci WM, Steger-May K, et al. Postoperative radiographic assessment of acetabular fractures: a comparison of plain radiographs and CT scans. J Orthop Trauma 2005;19:299–304.

[4] Keogh CF, Munk PL, Gee R, et al. Imaging of the painful hip arthroplasty. AJR Am J Roentgenol 2003;180:115–20.

[5] Petersilge CA. Chronic adult hip pain: MR arthrography of the hip. Radiographics 2000; 20:S43–52.

[6] Potter HG, Nestor BJ, Sofka CM, et al. Magnetic resonance imaging after total hip arthroplasty: evaluation of periprosthetic soft tissue. J Bone Joint Surg [Am] 2004;86-A:1947–54.

[7] Mintz DN, Hooper T, Connell D, et al. Magnetic resonance imaging of the hip: detection of labral and chondral abnormalities using noncontrast imaging. Arthroscopy 2005;21:385–93.

[8] Weissman BN. Imaging of total hip replacement. Radiology 1997;202:611–23.

[9] Heetveld MJ, Raaymakers EL, van Walsum AD, et al. Observer assessment of femoral neck radiographs after reduction and dynamic hip screw fixation. Arch Orthop Trauma Surg 2005;125: 160–5.

[10] Eustace S, Jara H, Goldberg R, et al. A comparison of conventional spin-echo and turbo spin-echo imaging of soft tissues adjacent to orthopedic hardware. AJR Am J Roentgenol 1998; 170:455–8.

[11] Herberts PG, Stromberg CN, Malchau H. Revision hip surgery: the challenge. In: Galante JO, Rosenberg AG, Callaghan JJ, editors. Total hip revision surgery: Bristol-Myers Squibb/Zimmer Orthopaedic Symposium series. New York: Raven; 1995. p. 1–15.

[12] NIH Consensus Conference. Total hip arthroplasty. JAMA 1995;273:1950–6.

[13] Knuesel PR, Pfirrman CWA, Noetzli HP, et al. MR arthrography of the hip: diagnostic performance of a dedicated water-excitation 3D double-echo steady-state sequence to detect cartilage lesions. AJR Am J Roentgenol 2004;183:1729–35.

[14] Leunig M, Beck M, Kalhor M, et al. Fibrocystic changes at anterosuperior femoral neck: prevalence in hips with femoroacetabular impingement. Radiology 2005;236:237–46.

[15] Hodler J, Yu JS, Goodwin D, et al. MR arthro-

graphy of the hip: improved imaging of the acetabular labrum with histologic correlation. AJR
Am J Roentgenol 1995;165:887–91.

[16] Czerny C, Hofmann S, Urban M, et al. MR arthrography of the adult acetabular capsular-labral
complex: correlation with surgery and anatomy.
AJR Am J Roentgenol 1999;173:345–9.

[17] Edwards DJ, Lomas D, Villar RH. Diagnosis of
the painful hip by magnetic resonance imaging
and arthroscopy. J Bone Joint Surg [Br] 1995;
77-B:374–6.

[18] Leunig M, Werlen S, Ungersbock A, et al. Evaluation of the acetabular labrum by MR arthrography. J Bone Joint Surg Br 1997;79-B:230–4.

[19] Petersilge CA, Haque MA, Petersilge WJ, et al.
Acetabular labral tears: evaluation with MR
arthrography. Radiology 1996;200:231–5.

[20] Czerny C, Hofmann S, Neuhold A, et al. Lesions
of the acetabular labrum: accuracy of MR
imaging and MR arthrography in detection and
staging. Radiology 1996;200:225–30.

[21] Schmid MR, Nötzli HP, Zanetti M, et al. Cartilage
lesions in the hip: diagnostic effectiveness of MR
arthrography. Radiology 2003;226:382–6.

[22] Fishman EK, Kuszyk B. 3D imaging: musculoskeletal applications. Crit Rev Diagn Imaging
2001;42:59–100.

[23] Tonnis D, Heinecke A. Acetabular and femoral
anteversion: relationship with osteoarthritis of
the hip. J Bone Joint Surg [Am] 1999;81-A:
1747–70.

[24] Sugano N, Ohzono K, Nishii T, et al. Computed-
tomography-based computer preoperative planning for total hip arthroplasty. Comput Aided
Surg 1998;3:320–4.

[25] Keene GS, Villar RN. Arthroscopic anatomy of
the hip: an in vivo study. Arthroscopy 1994;10:
392–9.

[26] Cotten A, Boutry N, Demondion X, et al.
Acetabular labrum: MRI in asymptomatic volun-
teers. J Comput Assist Tomogr 1998;22:1–7.

[27] Ferguson SJ, Bryant JT, Ganz R, et al. An in vitro
investigation of the acetabular labral seal in hip
joint mechanics. J Biomech 2003;36:171–8.

[28] Petersilge CA. MR arthrography for evaluation of
the acetabular labrum. Skeletal Radiol 2001;
30:423–30.

[29] Dinauer PA, Murphy KP, Carroll JF. Sublabral
sulcus at the posteroinferior acetabulum: a potential pitfall in MR arthrography diagnosis of
acetabular labral tears. AJR Am J Roentgenol
2004;183:1745–53.

[30] Plotz GM, Brossmann J, Schunke M, et al.
Magnetic resonance arthrography of the acetabular labrum. Macroscopic and histological correlation in 20 cadavers. J Bone Joint Surg Br
2000;82-B:426–32.

[31] Fitzgerald RH. Acetabular labrum tears. Clin
Orthop 1995;311:60–8.

[32] Abe I, Harada Y, Oinuma K, et al. Acetabular labrum: abnormal findings at MR imaging in
asymptomatic hips. Radiology 2000;216:576–81.

[33] Swiontkowski MF. Current concepts review:
Intracapsular fractures of the hip. J Bone Joint
Surg [Am] 1994;76-A:129–38.

[34] Gomez-Castresana F, Caballer AP, Portal LF.
Avascular necrosis of the femoral head after
femoral neck fractures. Clin Orthop 2002;399:
87–109.

[35] Parker MJ. The management of intracapsular
fractures of the proximal femur. J Bone Joint Surg
[Br] 2000;82-B:937–41.

[36] Bosch U, Schreiber T, Krettek C. Reduction and
fixation of intracapsular fractures of the proximal
femur. Clin Orthop 2002;399:59–71.

[37] Baumgaertner MR, Curtin SL, Lindskog DM,
et al. The value of the tip-apex distance in predicting failure of fixation of peritrochanteric
fractures of the hip. J Bone Joint Surg [Am] 1995;
77-A:1058–64.

[38] Scherl SA. Common lower extremity problems
in children. Pediatr Rev 2004;25:52–62.

[39] Goodman DA, Feighan JE, Smith AD, et al. Subclinical slipped capital femoral epiphysis. Relationship to osteoarthrosis of the hip. J Bone Joint
Surg [Am] 1997;79-A:1489–97.

[40] Wenger DE, Kendell KR, Miner MR, et al. Acetabular labral tears rarely occur in the absence
of bony abnormalities. Clin Orthop 2004;426:
145–50.

[41] Hungerford DS. Role of core decompression as
treatement method for ischemic femur head
necrosis. Orthopade 1990;19:219–23.

[42] Wirtz C, Zilkens KW, Adam G, et al. MRI-
controlled outcome after core decompression of
the femur head in aseptic osteonecrosis and
transient bone marrow edema. Z Orthop 1998;
136:138–46.

[43] Stulberg BN, Davis AW, Bauer TW, et al. Osteonecrosis of the femoral head. A prospective randomized treatment protocol. Clin Orthop 1991;
268:140–51.

[44] Beltran J, Knight CT, Zuelzer WA, et al. Core decompression for avascular necrosis of the femoral head: correlation between long-term results
and pre-operative MR staging. Radiology 1990;
175:533–6.

[45] Radke S, Kirschner S, Seipel V, et al. Magnetic
resonance imaging criteria of successful core decompression in avascular necrosis of the hip.
Skeletal Radiol 2004;33:519–23.

[46] Manaster BJ. Total hip arthroplasty: radiographic
evaluation. Radiographics 1996;16:645–60.

[47] Kraemer WJ, Soplys R, Waddell JP, et al. Bone
scan, gallium scan and hip aspiration in the
diagnosis of infected total hip arthroplasty.
J Arthroplasty 1993;8:611–5.

[48] Roberts P, Walters AJ, McMinn DJ. Diagnosing
infection in hip replacement: the use of fine
needle aspiration and radiometric culture. J Bone
Joint Surg [Br] 1992;74-B:265–9.

[49] Maus TP, Berquist TH, Bender CE, et al. Arthrographic study of painful total hip arthroplasty:
refined criteria. Radiology 1987;162:721–7.

[50] Petersilge CA, Lewin JS, Duerk JL, et al. Optimizing imaging parameters for MR evaluation of the spine with titanium pedicle screws. AJR Am J Roentgenol 1996;166:1213–8.

[51] Potter HG, Montgomery KD, Padgett DE, et al. Magnetic resonance imaging of the pelvis: new orthopaedic applications. Clin Orthop 1995;19:223–31.

[52] Tartaglino LM, Flanders AE, Vinitski S, et al. Metallic artifacts on MR images of the postoperative spine: reduction with fast spin-echo techniques. Radiology 1994;190:565–9.

[53] Tormanen J, Tervonen O, Koivula A, et al. Image technique optimization in MR imaging of a titanium alloy joint prosthesis. J Magn Reson Imaging 1996;6:805–11.

[54] Sugimoto H, Hirose I, Miyaoka E, et al. Low-field-strength MR imaging of failed hip arthroplasty: association of femoral periprosthetic signal intensity with radiographic, surgical, and pathologic findings. Radiology 2003;229:718–23.

[55] Lemmens JA, van Horn JR, den Boer J, et al. MR imaging of 22 Charnley- Muller total hip prostheses. ROFO Fortschr Geb Rontgenstr Nuklearmed 1986;145:311–5.

[56] Weiland DE, Walde TA, Leung SB, et al. Magnetic resonance imaging in the evaluation of periprosthetic acetabular osteolysis: a cadaveric study. J Orthop Res 2005;23:713–9.

[57] Matsuura H, Inoue T, Konno H, et al. Quantification of susceptibility artifacts produced on high-field magnetic resonance images by various biomaterials used for neurosurgical implants. Technical note. J Neurosurg 2002;97:1472–5.

[58] Kolind SH, MacKay AL, Munk PL, et al. Quantitative evaluation of metal artifact reduction techniques. J Magn Reson Imaging 2004;20:487–95.

[59] Li T, Mirowitz SA. Fast T2-weighted MR imaging: impact of variation in pulse sequence parameters on image quality and artifacts. Magn Reson Imaging 2003;21:745–53.

[60] Cho ZH, Kim DJ, Kim YK. Total inhomogeneity correction including chemical shifts and susceptibility by view angle tilting. Med Phys 1988;15:7–11.

[61] Butts K, Pauly JM, Gold GE. Reduction of blurring in view angle tilting MRI. Magn Reson Med 2005;53:418–24.

[62] Pfirrmann CW, Notzli HP, Dora C, et al. Abductor tendons and muscles assessed at MR imaging after total hip arthroplasty in asymptomatic and symptomatic patients. Radiology 2005;235:969–76.

[63] Masonis JL, Bourne RB. Surgical approach, abductor function, and total hip arthroplasty dislocation. Clin Orthop 2002;405:46–53.

[64] Kwong LM, Jasty M, Mulroy RD, et al. The histology of the radiolucent line. J Bone Joint Surg [Br] 1992;74-B:67–73.

[65] Pfahler M, Schidlo C, Refior HJ. Evaluation of imaging in loosening of hip arthroplasty in 326 consecutive cases. Arch Orthop Trauma Surg 1998;117:205–7.

[66] Eustace S, Shah B, Mason M. Imaging orthopedic hardware with an emphasis on hip prostheses. Orthop Clin North Am 1998;29:67–84.

[67] Palestro CJ, Brown ML, Forstrom LA, et al. Society of Nuclear Medicine procedure guideline for ^{111}In-leukocyte scintigraphy for suspected infection/inflammation. Reston (VA): Society of Nuclear Medicine; 1997. p. 75–8.

[68] Weissman BN. Imaging the total hip replacement. Radiology 1997;202:611–23.

[69] Anthony PP, Gie GA, Howie CR, et al. Localized endosteal bone lysis in relation to the femoral components of cemented total hip arthroplasties. J Bone Joint Surg [Br] 1990;72:971–9.

[70] White LM, Kim JE, Mehta M, et al. Complications of total hip arthroplasty: MR imaging—initial experience. Radiology 2000;215:254–62.

[71] Bogost GA, Lizerbram EK, Crues JV. MR imaging in evaluation of suspected hip fracture: frequency of unsuspected bone and soft-tissue injury. Radiology 1995;197:263–7.

[72] Mintz D, Hooper T, Connell D, et al. Magnetic resonance imaging of the hip: detection of labral and chondral abnormalities using noncontrast imaging. Arthroscopy 2005;21:385–93.

[73] The classic: arthroplasty of the hip: a new operation by John Charnley, M.B., B. Sc. Manc., F.R.C.S. Reprinted from Lancet p. 1129–32, 1961. Clin Orthop Relat Res 1973;Sep:4–8.

RADIOLOGIC
CLINICS
OF NORTH AMERICA

Radiol Clin N Am 44 (2006) 367–389

Postoperative Imaging of the Knee

Matthew A. Frick, MD*, Mark S. Collins, MD, Mark C. Adkins, MD

- Postoperative imaging
- Anterior cruciate ligament
- Posterior cruciate ligament
- Collateral ligaments
- Menisci

- Knee arthroplasty (total and unicompartmental)
- Articular cartilage
- References

The knee is one of the most frequently injured joints in the body, particularly in young athletes, and is a common site of age-related degenerative change. More and more commonly, many knee injuries are being treated operatively, which has led to an increase in the number of postoperative cases radiologists encounter in their daily practice. Accordingly, this topic recently has been given much attention in the radiology literature and there are several authoritative and exhaustive reviews of this topic [1–8]. Yoshida and Recht authored a thorough and in-depth review of postoperative knee imaging in *Radiologic Clinics of North America* in 2002 [190]. This manuscript reviews recent advances in operative techniques and the attendant imaging findings, which have evolved since their report.

Knee pathology leading to surgery most often is the result of age-related degenerative change or sports-related injuries. Total knee arthroplasty (TKA), anterior cruciate ligament (ACL) reconstruction, and meniscal repair have been, and continue to be, the surgeries performed most commonly; however, primary repair of chondral injuries is performed with increasing frequency. Radiologists should be familiar with these procedures, the expected postoperative imaging appearance unique to each procedure, and the appearance of potential postoperative complications.

Although patients may present for routine postoperative imaging after successful surgery, the majority of patients who present for imaging in the postoperative period typically are experiencing new or persistent symptoms, such as pain, decreased range of motion, or instability. The goal of imaging in these patients is to exclude a postoperative complication or the development of metachronous pathology. Knowledge of the surgery undertaken and the expected imaging findings at various stages of healing for each surgery constitute the context in which imaging studies should be evaluated. It is recommended that, whenever possible, any available operative notes should be reviewed before imaging.

Postoperative imaging

Radiologists should be aware of general and operation- or procedure-specific issues. After any knee surgery, it is routine surgical practice to obtain immediate postoperative radiographs to exclude the presence of iatrogenic foreign bodies, such as sponges, needles, or other instruments. In the subacute postoperative period, the most common general postoperative imaging considerations include hemorrhage and infection. In the knee joint, both

Department of Radiology, Division of Musculoskeletal Radiology, Mayo Clinic and Foundation, Rochester, MN, USA
* Corresponding author. Division of Musculoskeletal Radiology, Department of Radiology, Mayo Clinic and Foundation, 200 First Street, SW, Rochester, MN 55905.
E-mail address: frick.matthew@mayo.edu (M.A. Frick).

0033-8389/06/$ – see front matter © 2006 Elsevier Inc. All rights reserved.
radiologic.theclinics.com

doi:10.1016/j.rcl.2006.02.001

may manifest clinically with localized pain and swelling at the operative site. Radiographically, soft tissue swelling may be evident along with the presence of a knee effusion. Clinical and serologic indicators (ie, fever, white blood cell [WBC] count, erythrocyte sedimentation rate, and C-reactive protein) generally allow differentiation of these entities, but joint aspiration may be necessary in enigmatic cases.

Radiologic evaluation of the postoperative knee should, in all cases, begin with routine radiographic projections (anteroposterior [AP], lateral and patellar sunrise, or Merchant views). Notch views also may be beneficial in some cases. In some cases, radiographic examination is diagnostic. In many cases, however, additional imaging ise necessary. Although CT and MR imaging once were believed of limited value in evaluating the postoperative knee because of the imaging-related artifacts unique to each modality (ie, beam hardening and metallic susceptibility artifacts, respectively), recent advances in scanner technology and software have made great strides in eliminating or reducing these artifacts, bringing these modalities to the forefront of postoperative knee imaging. At the same time, historically important modalities, such as arthrography and scintigraphy, play a less central role in postoperative knee imaging than in the past. Several investigators recently reviewed the usefulness and importance of each of these modalities in the assessment of the postoperative knee [4,6,7,9].

Anterior cruciate ligament

The ACL is injured frequently, particularly in young athletic individuals, and is the ligament in the knee repaired most commonly [10]. Decisions regarding therapy are predicated by several factors, including the degree of ACL disruption (partial versus complete), patient age, and patient activity level. Most orthopedic surgeons offer conservative treatment in the setting of partial tears or for those patients who have a sedentary, nonactive lifestyle but opt to repair complete disruptions, particularly in young, active patients, because of the known long-term sequelae of ACL-deficient knees [11–15].

The preferred method of treatment for ACL disruption is ligament reconstruction, usually arthroscopically guided. Primary repair of the more common midsubstance tears is associated with poor long-term outcomes. One important exception is avulsive type injuries of the ACL that, although uncommon, may be seen occasionally in children. These injuries tend to occur at the femoral origin or tibial insertion and may be repaired primarily with good outcome.

Reconstruction options include tendon autografts, allografts, and synthetic grafts (Dacron, Gore-Tex, carbon fiber, and so forth). The latter are associated with the development of intra-articular debris and sterile effusions and, therefore, no longer are used [16]. Cadaveric allografts may be advantageous in that there is no associated donor site morbidity; however, there is at least a theoretic risk for disease transmission [5].

ACL reconstructions usually are performed with bone-patellar tendon-bone (BPTB) or hamstring (distal semimembranosus and gracilis) autografts. Although BPTB autografts remain the most common method of reconstruction, recent investigations indicate similar long-term success with hamstring autografts [17–20]. In particular, it was believed in the past that patellar tendon autografts were superior in young, active athletes as the bone plug at either end of the harvested graft allowed for superior fixation and an earlier return to competition [21]. More recent work, however, suggests that hamstring autografts may be superior, even in high-level athletes [22]. Although an entire treatise could entertain the attributes of each surgical technique, Ilasan and colleagues offer a concise summary of the major advantages and disadvantages of the BPTB and hamstring autografts and state that ultimately the choice depends on surgeon preference and training [5]. The quadriceps tendon and iliotibial band tendon are less frequent donor sites for autograft reconstruction.

The goal of the reconstructed ACL is to provide isometry throughout the course of knee flexion and extension [5]. The location of the femoral and tibial tunnels and graft position are critical to achieving isometry and play a central role in the avoidance of impingement and, thus, the overall long-term success and viability of the graft [Fig. 1] [23–27]. In general, a well-positioned femoral tunnel is the most important factor for obtaining isometry, and a well-positioned tibial tunnel is the most important factor for avoidance of impingement.

The anterior gin of the tibial tunnel should lie posterior to the point where extension of Blumenstaat's line intersects the tibia but should not lie posterior to the midpoint of the proximal tibia. When the tunnel is located too far anteriorly, patients may experience limitation of terminal extension secondary to roof impingement, and when the tunnel lies too far posteriorly, patients often are beset with persistent instability [7,23]. Similarly, the femoral tunnel should lay posterior to the intersection of Blumenstaat's line and a line parallel to the posterior femoral cortex. The size of the intercondylar notch also is a critical factor in the development of graft impingement. The smaller the notch, the more likely that sidewall

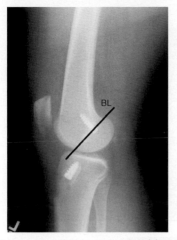

Fig. 1. ACL reconstruction in a 20-year-old woman. The lateral projection is taken with patient in full extension, which is necessary to evaluate tunnel position adequately. The femoral tunnel should lie posterior to the intersection of Blumenstaat's line (BL) with a line parallel to the posterior femoral cortex. The tibial tunnel should lie posterior to the intersection of Blumenstaat's line with the tibial plateau but not posterior to the midpoint of the tibia.

impingement occurs. This latter problem, however, is remedied easily with notchplasty at the time of reconstruction [28–30].

Radiographs offer valuable information about the positioning of the femoral and tibial tunnels and the location of the titanium interference screws used for fixation of the bone plugs at either end of the BPTB autograft. Unfortunately, aside from assessment of tunnel location and position of the tibia relative to the femur, little information regarding the integrity of the graft itself may be gained from radiographs. Recent advances in scanner technology and software have allowed MR imaging to become the preferred modality for evaluating postoperative ACL.

The expected MR imaging appearance of an ACL reconstruction varies depending on the type of graft used and, more importantly, on the timing of imaging relative to graft placement [6,7,31–35]. In the immediate (<1 month) postoperative period, the graft typically demonstrates low T1 and T2 signal intensity, similar to the native patellar tendon [24,36]. This imaging appearance is attributed to the avascular nature of the graft during this time period [7]. After 1 month and up to 1 to 2 years after surgery, there may be increased T2 signal intensity within the graft as it undergoes a process of revascularization and resynovialization, a phenomenon referred to as graft "ligamentization" [7]. Although the graft is vulnerable to injury during

this period, increased signal intensity should not be mistaken for evidence of a tear unless the increased signal intensity is equal to that of fluid [30]. By 2 years, the graft should be uniformly low in signal intensity on all imaging sequences, much as the native ACL appears [Fig. 2] [37].

Known complications of ACL reconstruction include graft failure, graft impingement, and arthrofibrosis [6,24,26,35,38–42]. In a recent series of 90 patients, the most common cause of a failed ACL reconstruction was surgical technical errors followed by traumatic reinjuries [43]. Septic arthritis after ACL reconstruction is an exceedingly rare complication, occurring in less than 0.5% of patients [38]. Resorption of the bone plug is reported as a cause of ACL graft failure, but this also is an uncommon complication [44].

The bone plugs at either end of a BPTB autograft may be secured within the femoral and tibial tunnels with titanium or bioabsorbable interference screws. Although bioabsorbable screws offer the advantage of eliminating the metallic susceptibility artifact seen on MR imaging with metallic and titanium screws, they have a few important disadvantages. Bioabsorbable screws are associated with inflammatory cell-mediated foreign body reactions, which may be associated with failure of fixation [45,46]. In rare cases, the screws may egress completely and become dislodged within the knee joint [Fig. 3]. For these reasons, the orthopedic surgeons at the authors' institution use titanium interference screws most commonly. Endobuttons (Acufex Microsurgical, Mansfield, Massachusets) are an effective alternative to interference screws

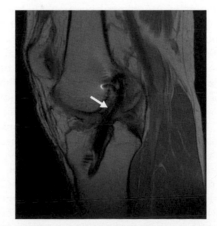

Fig. 2. Normal ACL BPTB allograft in a 28-year-old man. Sagittal FSE proton density MR image demonstrates a well-positioned, intact ACL BPTB autograft. Note a few streaks of increased signal intensity parallel to the fibers of the graft (*arrow*), which occasionally may be seen.

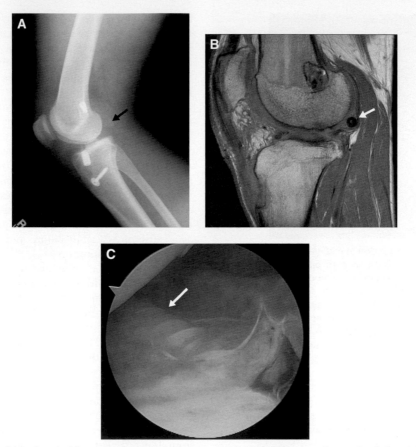

Fig. 3. Displaced bioabsorbable screw from ACL reconstruction. (A) Lateral radiograph of the right knee of a 20-year-old man who had undergone prior ACL reconstruction, and subsequent revision demonstrates titanium interference screws in the femoral and tibial tunnels. There is a small circular density posterior to the knee (arrow). (B) Sagittal proton density MR images demonstrate a displaced bioabsorbable screw within the posterior recess of the joint (arrow). (C) At arthroscopy, the egressed screw was embedded in fibrinous material (arrow).

in fixation of the femoral component; however, there use is associated with increased risk for cystic degeneration of the graft [47,48]. Some investigators also have examined the role of reconstructing the anterolateral and posteromedial bands of the ACL separately in a two-bundle technique with the use of endobutton fixation [49]. This technique is not in common practice at the authors' institution.

On MR imaging, impinged grafts demonstrate increased T2-weighted signal intensity, especially within the distal two thirds of the graft [Fig. 4]. Although this imaging finding also may be seen with normal evolution of the ACL graft (discussed previously), abnormal signal intensity resulting from impingement does not resolve but, rather, worsens, with time. In the setting of impingement, surgeons may elect to perform notchplasty to elevate the roof or widen the notch, depending on the factor believed responsible for the impingement. After notchplasty, abnormal signal intensity in the graft should resolve within 12 weeks [28,50].

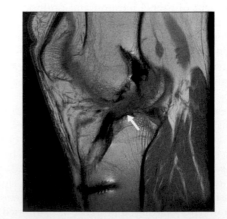

Fig. 4. Suboptimal tibial tunnel placement for ACL graft in a 39-year-old man. Sagittal FSE proton density (2000/18) MR image demonstrates postoperative changes of ACL reconstruction with increased signal intensity in the mid and distal graft (arrow) secondary to impingement. Note the position of the tibial tunnel is too far anterior (anterior to the expected intersection of Blumenstaat's line).

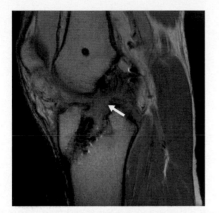

Fig. 5. ACL graft rupture. Sagittal FSE proton density (3000/26) MR image demonstrates postoperative changes of ACL reconstruction with complete disruption of the midsubstance of the graft (*arrow*), consistent with recurrent tear. A completely disrupted graft was confirmed at revision surgery.

ACL grafts may fail at any time but are vulnerable particularly during the remodeling process. As with the native ACL, graft failure may be detected or suggested by direct or indirect findings seen on MR imaging. Anterior translation of the tibia relative to the femur, the so-called "MR anterior-drawer sign," suggests failure or laxity of the graft and should be noted in the report, thereby allowing clinicians to correlate this finding with those at physical examination.

With conventional MR imaging after ACL reconstruction, T2-weighted images play a central role in determining graft integrity. Failure to confidently identify graft fibers extending from the femoral tunnel to the tibial tunnel or the presence of increased signal intensity (equal to that of fluid) in the expected location of the graft should be taken as evidence of disruption [Fig. 5]. Some investigators show that this finding is appreciated better at MR arthrography [6]. McCauley and colleagues demonstrate a sensitivity of 100% and specificities of 89% to 100% when using MR arthrography for evaluation of graft failure [35]. The authors typically image the post-ACL reconstruction knee with conventional MR imaging sequences and reserve MR arthrography for indeterminate or equivocal cases.

Localized anterior arthrofibrosis is described as a complication of ACL reconstruction [38,40–42]. Histopathologically, this process represents synovial hyperplasia with excessive production of fibrous tissue and inflammatory cell infiltration [5]. Colloquially, this finding is referred to as the cyclops lesion, owing to its appearance at arthroscopy [41]. Although the exact origin of this process is unclear, what is clear is that the tissue may cause a mechanical limitation to full extension and also

may be a cause of knee pain at extension [6,51]. MR imaging typically demonstrates an area of low T1- and low to intermediate T2-weighted signal intensity within the soft tissues anterior to the graft and cephalad to the tibia [Fig. 6] [41,42]. When present, areas of increased T2-weighted signal intensity within these lesions are believed the result of inflammatory cell infiltrates seen histologically. Surgical resection is curative.

A potential late complication of ACL grafts that may be seen on MR imaging is the development of ganglia, which represent cystic degeneration of the graft. Ganglia frequently are associated with bone tunnel enlargement and appear similar to ganglia within the native ACL [38]. This finding is not associated with graft failure or instability.

Donor site complications are more common with BPTB autografts. Controversy exists as to whether or not the defect within the native patellar tendon should be closed at the time of operation or be allowed to heal spontaneously. Although some investigators favor the latter approach, it is common practice at the authors' institution to primarily close the harvest site defect [52]. In the early postoperative period, MR imaging may demonstrate loss of tendon definition and increased signal intensity within a diffusely thickened patellar tendon, which generally resolves by 12 to 18 months postoperatively [5]. Patellar fracture and patellar tendon rupture are described but fortunately are rare [Fig. 7] [53–55]. Similarly, harvesting a BPTB graft of up to 10 mm in width has no adverse consequences on quadriceps/extensor mechanism strength [56].

Bone tunnel enlargement is reported after ACL reconstruction surgery [Fig. 8] [57]. The importance of this phenomenon is not yet known. Tun-

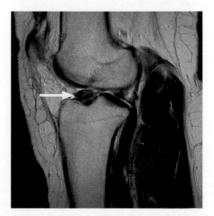

Fig. 6. Cyclops lesion. Sagittal T2-weighted image of the right knee demonstrates an arthrofibrotic nodule within Hoffa's fat pad (*arrow*) consistent with a cyclops lesion.

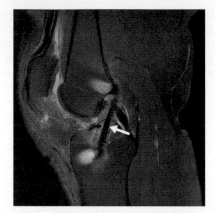

Fig. 7. Patellar tendon rupture after ACL reconstruction. Sagittal fat-saturated T2-weighted MR image demonstrates rupture of the right patellar tendon, which occurred during a fall down stairs 2 days after ACL reconstruction with a BPTB autograft. Note the ACL graft is intact (*arrow*).

nel lysis or expansion may be clinically significant in revision surgery, because the enlarged tunnels may complicate graft placement and fixation. Although many theories exist as to the development of this phenomenon, the cause most likely is multifactorial [58]. Recent evidence suggests a biomechanical basis, but further studies are needed on this issue [59].

Posterior cruciate ligament

In the past, orthopedic surgeons typically observed patients who had isolated posterior cruciate ligament (PCL) injuries. Evolution of the understand-

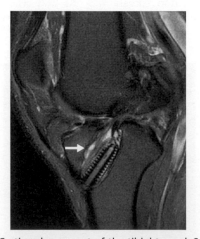

Fig. 8. Cystic enlargement of the tibial tunnel. Sagittal fat-saturated T2-weighted MR image demonstrates cystic enlargement of the tibial tunnel (*arrows*) after ACL reconstruction. This patient was not believed to have instability on clinical examiantion and is being followed.

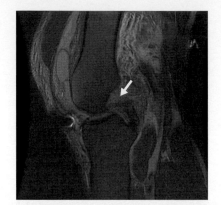

Fig. 9. PCL rupture in a 32-year-old man. Sagittal fat-saturated T2-weighted MR image demonstrates marked thickening and increased signal intensity within the proximal PCL (*arrow*), consistent with a high-grade injury. At surgery, complete disruption of the PCL was confirmed and the PCL was reconstructed with an Achilles tendon autograft.

ing of the PCL-deficient knee, however, has led to increased interest in primary repair of this structure [60,61]. Today, most orthopedic surgeons continue to treat the less severe, grade 1 and grade 2 injuries nonoperatively but repair the more serious, grade 3 lesions [Fig. 9]. These latter injuries occur commonly in association with other ligamentous and soft tissue injuries in the knee that need repair [62].

Two different surgical techniques may used to reconstruct the PCL—the tibial inlay procedure or the two-bone tunnel procedure [63–65]. The tibial inlay procedure usually is performed with a BPTB autograft and the latter, the two-bone tunnel procedure, may be performed with hamstring or BPTB autograft material or with allograft.

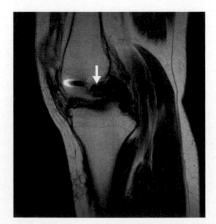

Fig. 10. Intact PCL graft. Sagittal T2-weighted image without fat saturation demonstrates an intact PCL graft (*arrow*), reconstructed with the two-bone tunnel patellar tendon autograft technique.

MR imaging clearly is the imaging study of choice in evaluating a PCL reconstruction; however, little exists in the literature on this topic. In one study, 20 patients were evaluated between 1 and 2 years postoperatively [66]. In this series, the predominant finding was uniform low signal intensity throughout the graft [Fig. 10]. A slightly smaller series, however, demonstrates mild to moderate increased T1- and T2-weighted signal intensity within thickened PCL grafts during the first postoperative year [63]. This finding is akin to the findings seen within ACL grafts during this period and likely is the result of similar histologic evolution of the graft. Thus, it likely is normal to see a small amount of increased signal intensity within a PCL graft within the first 2 years postoperatively, but after this, the graft should be uniformly dark. Arthrofibrotic tissue also may be seen anterior to a PCL graft and may, in rare cases, require resection [67,68].

Collateral ligaments

Isolated collateral ligament injuries usually are treated conservatively. As with the PCL, higher-grade (grade 3) injuries may be repaired surgically. Primary repair is the treatment of choice for these injuries. The lateral collateral ligament or fibular collateral ligament may be repaired with an allograft or autograft (Achilles tendon, BPTB, hamstring tendon, biceps femoris tendon, or iliotibial tract) [69–72]. Controversy exists as to when and how the medial collateral ligament should be repaired [73].

Surgical repair of the collateral ligaments may require the use of metallic hardware, which may cause degradation of image quality on MR imaging resulting from susceptibility artifact. Allowing for this limitation, however, MR imaging still is the imaging study of choice for postoperative evaluation of collateral ligament integrity. As with native collateral ligaments, laxity of the fibers or the presence of increased signal intensity (equal to that of fluid) interrupting the ligament should be taken as evidence of recurrent disruption.

Menisci

Tears of the menisci are one of the pathologies encountered most frequently in the knee and may occur with age-related degeneration, trauma, or a combination of the two. In the past, surgical repair of the meniscus entailed simply complete open removal of the torn or degenerated meniscus. Longitudinal studies, however, bear witness to the development of advanced, precocious osteoarthritis in these patients [74–77]. Currently, a torn or de-

generated meniscus may be addressed surgically in one of three ways—partial meniscectomy, primary repair, or replacement with allograft transplantation.

Current understanding of the biomechanics of the menisci suggest that in addition to providing stability across the knee joint, the menisci more importantly act to dissipate the transmission of forces from the femur to the tibia by effectively increasing the contact area [77,78]. The peripheral third of the meniscus, composed of circumferentially oriented fibers, plays a critical role in maintaining hoop stress during mechanical loading [79]. Magee and colleagues recently have shown an increased incidence of radial tears in knees that underwent prior partial meniscectomy, which they attribute to a decreased ability of the meniscus to distribute forces placed on it evenly [80].

Given that an intact blood supply is necessary for healing, an additional and equally important histologic feature of the peripheral third of the meniscus is its communication with the perimeniscal vascular plexus [7]. It generally is believed that those lesions occurring in the peripheral one third of the meniscus (so-called "red-white" tears) are most amenable to primary repair. Lesions located within the central, avascular two thirds of the meniscus (so called "white-white" tears), however, are less likely to heal primarily and, therefore, typically are treated with partial meniscectomy. Regardless of the operative intervention, one of the most important determinants of short- and long-term success it that the postsurgical meniscus resembles, as closely as possible, the normal triangular, semilunar morphology of the native, uninjured meniscus [75].

The challenge then, for orthopedists, is to retain as much of the meniscus as possible by way of primary repair or limited partial meniscectomy, removing only the unstable tissue [1,76,81]. In essence, a circumferential-type partial meniscectomy is preferable to a segmental partial meniscectomy [1]. Meniscal repair, performed less often than partial meniscectomy, preserves the meniscus but is associated with a higher short-term complication rate [6].

Some meniscal tears may be treated nonoperatively. Stable, horizontally oriented tears and longitudinal tears extending less than 1 cm that are stable at arthroscopy may be observed [82,83]. This is true particularly if these tears occur in the outer, or vascular, third of the meniscus, where spontaneous healing may be expected [84]. Unstable meniscal defects may be repaired with sutures, using either an inside-out or outside-in technique, or fixed with a variety of devices on the market, including bioabsorbable arrows, tacks, or darts [1]. Healing of meniscal repair typically

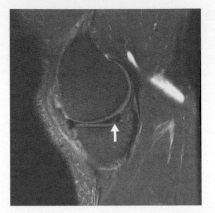

Fig. 11. Meniscal retear in a 35-year-old man. Sagittal, fat-saturated, T2-weighted MR image demonstrates a recurrent tear within the posterior horn of the medial meniscus (*arrow*), which was confirmed arthroscopically.

occurs by 4 months and correlates well with clinical symptoms [1].

Because postoperative changes may mimic or obscure tears, MR imaging initially was believed unreliable in assessing the postoperative meniscus

[85–91]. Evolution of the understanding of the expected appearance of the postoperative meniscus, however, has allowed MR imaging to become the imaging modality of choice for assessing the post-surgical meniscus. The widely accepted diagnostic criteria of a tear in the native meniscus (ie, increased signal intensity extending to an articular surface or abnormal, irregular morphology on short echo time MR images) are, alone, unreliable in detection of a torn meniscus in the postoperative state unless these findings are distant from the operative site or there is clear fragmentation or displaced meniscal fragments [see **Figs. 11 and 12**] [7,92,93]. When conventional MR imaging sequences are used to evaluate the postoperative meniscus, high-quality, high-contrast resolution, T2-weighted sequences, preferably with fat saturation, are of paramount importance. The presence of increased signal intensity, equal to that of synovial fluid, within a linear defect in the meniscus extending to at least one articular surface or displaced meniscal fragments is suggestive of recurrent tear. The former finding offers high sensitivity (88%–92%) but low specificity (41%–69%) [86,89,90].

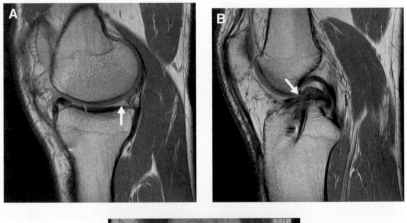

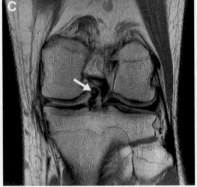

Fig. 12. Recurrent bucket-handle tear of the medial meniscus. A 21-year-old man who previously underwent prior BTPB allograft reconstruction of the ACL and repair of a bucket-handle tear of the medial meniscus. Sagittal FSE proton density (*A, B*) and coronal fat-saturated T2-weighted (*C*) MR images demonstrate a large, recurrent bucket-handle tear of the medial meniscus that is displaced into the intercondylar notch (*arrows*).

Whether or not conventional MR imaging alone can assess the postoperative meniscus accurately remains controversial. Currently there are two major approaches to MR imaging of the postoperative meniscus. Some invesetigators advocate beginning with conventional MR and proceeding to MR arthrography when the former is equivocal or inconclusive, especially if the repair involves less than 25% of the meniscus [1]. McCauley, however, advocates the use of MR arthrography in all patients who have known prior meniscal intervention [6].

There are two advantages of MR arthrography over conventional MR imaging postoperatively. First and foremost is the benefit of joint distention, which forces contrast material into any persistent or recurrent meniscal tears or defects. The second advantage over conventional MR imaging is the increased contrast resolution of contrast material compared with synovial fluid [1]. The major disadvantage of MR arthrography is that it is invasive, exposing patients to the low but possible risk for bleeding or infectious complications. A large meta-analysis of studies related to MR arthrography, however, shows the procedure to be safe with low risk for complication [94]. Several comparative studies demonstrate improved accuracy for detecting recurrent meniscal tears when MR arthrography is compared with conventional MR imaging; however, the largest of these studies fails to show a statistically significant difference [86,95–97].

For MR arthrography, most investigators advocate the infusion of approximately 10 to 20 mL of a 1:200 dilution of saline and gadolinium (0.001 mmol/mL) into the knee joint using fluoroscopic or ultrasound guidance. A small amount of epinephrine may be added to the injectate to delay synovial resorption of the contrast media [1]. The imaging examination should consist, at minimum, of fat-saturated T1-weighted images in all three planes, one fat-saturated T2-weighted sequence, and at least one T1-weighted sequence without fat saturation [1].

Indirect MR arthrography, consisting of delayed imaging (10–20 minutes) after infusion of intravenous gadolinium and patient exercise, is shown, by some investigators, to offer diagnostic accuracy similar to that of direct MR arthrography in the assessment of recurrent or residual meniscal tears [7,97,98]. Lack of joint distention and enhancing granulation tissue may lead to a false-positive interpretation with this technique [1].

CT arthrography similarly is highly sensitive (100%) but nonspecific (78%) in demonstrating residual or recurrent meniscal tears postoperatively [99]. When the same data are reanalyzed using more stringent criteria, the sensitivity falls to 93%

but specificity increases to 89% [99]. Because of the advantage of being able to evaluate the adjacent soft tissues and bone marrow with MR imaging, CT arthrography typically is reserved for patients who have a contraindication to MR or for highly tailored clinical situations.

Meniscal transplantation was performed first in 1984 and is performed with increasing frequency. Transplantation typically is performed in younger, symptomatic patients who have irreparable tears or have failed prior partial meniscectomy [100]. As described by Toms and colleagues, meniscal allografts are transplanted with the anterior and posterior horns intact, attached to bone plugs or a bone bar placed in drill tracts and fixed with traction sutures [1]. The allograft alternatively may be fixed directly to bone with soft tissue anchors. Early long-term follow-up in this cohort of patients demonstrates a favorable outcome with deep-frozen meniscal allografts, similar to an intact meniscus, but shows poor longevity in lyophilized meniscal transplants, akin to a meniscectomy [101]. A more recent study suggests good short-term (mean of 40 months) outcomes [102]. The few reports in the literature regarding postoperative allograft imaging suggest that signs of failure include fragmentation or extrusion of the meniscus and progressive loss of the adjacent articular cartilage [Fig. 13] [7,103–105].

Osteonecrosis is a rare but potential complication after meniscal intervention with which radiologists should be familiar. This finding is described mainly in case reports [106,107]. Although it is

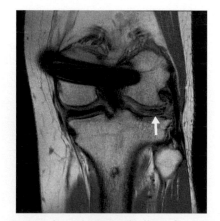

Fig. 13. Torn lateral meniscal allograft. A 24-year-old woman underwent prior ACL reconstruction and placement of a lateral meniscal allograft after failed partial meniscectomy. Coronal proton density MR image demonstrates a complex tear of the posterior horn of the meniscal allograft (*arrow*). Allograft tear was confirmed surgically and treated with partial meniscectomy. Metallic artifact is the result of prior distal femoral varus osteotomy.

believed that this complication may occur more often in elderly patients, the underlying pathophysiology is not well understood and warrants further investigation.

Knee arthroplasty (total and unicompartmental)

Miller recently authored an exhaustive review of postoperative TKA imaging to which little can be added nor can it be stated better [4]. Much of this discussion attempts to summarize the information from his work, to which readers are referred, and which is believed most salient to general radiologists.

Only the hip joint is replaced more often than the knee [108]. More than 350,000 primary total knee replacements and 29,000 revision procedures were performed in 2002 [109]. The most common indications for TKA are degenerative osteoarthritis and rheumatoid arthritis [4]. Although TKA is performed more frequently than unicompartmental arthroplasty, the latter may be undertaken when pathology is limited to a single compartment or there is an intact ACL or correctable knee alignment [Fig. 14] [110,111].

The most common total knee replacement systems use metal (cobalt-chromium) condylar and tibial components with a layer of polyethylene overlying the tibial component. Newer systems have introduced the use of titanium and zirconium. In all cases, the ACL is sacrificed with or without sacrifice of the PCL. All devices are semiconstrained or unconstrained, depending on the degree of stability [4]. Although controversy exists over whether or not the patella should be resurfaced, a recent, large meta-analysis of 10 studies suggests that patellar resurfacing reduces the risk for reoperation and the risk for persistent anterior knee pain [112–114]. It is common practice among the orthopedic surgeons at the authors' institution to resurface the patella with a polyethylene component cemented directly onto the prepared native patella.

With ideal implantation, the expected longevity of a TKA system is 10 to 15 years in most patients, with reports of up to 20 or more years in some series [Fig. 15] [115–118]. Survival times for unicompartmental systems are less than for total knee systems. Because the former often is used as a bridge to TKA, however, uncompartment prostheses may be converted to total joint arthroplasties if, and when, they fail. Patients who have complications after TKA generally present with nonspecific pain, and there is little insight to the cause of their symptoms. Imaging evaluation of the painful TKA should begin with routine radiographs [119].

TKA should reproduce normal anatomic alignment (5°–8° valgus) and be fitted to maintain proper tension and balancing of adjacent soft tissue structures [4]. The condylar component should mirror the size of the native femoral condyles and the anterior aspect should sit flush against and be parallel to the anterior cortex of the distal femur [4]. Oversized condylar components may cause mechanical limitation of movement, whereas undersized condylar components may result in instability and, ultimately, notching of the anterior femoral cortex, a condition that may predispose to fracture [4,120]. The tibial component should mirror the size of the native tibial plateau and should

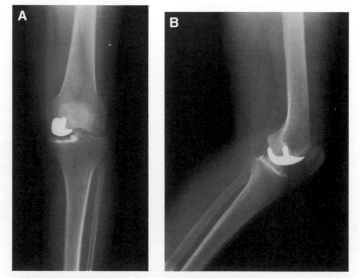

Fig. 14. Unicompartmental knee arthroplasty in a 67-year-old woman. AP (*A*) and lateral (*B*) radiographs of the left knee demonstrate a well-seated medial unicompartmental arthroplasty.

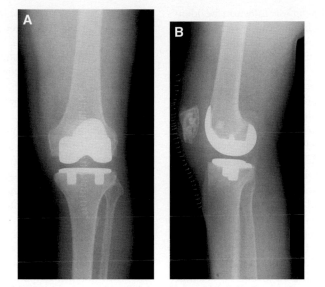

Fig. 15. TKA in a 62-year-old man. Postoperative AP (*A*) and lateral (*B*) radiographs demonstrate a well-seated TKA with patellar resurfacing.

be placed perpendicular to the tibial shaft on the AP projection with slight posterior-inferior tilt on the lateral view [4]. The height of the patella, from the top edge of the polyethylene component, should be the same as from the native tibial plateau and the total thickness of the patellar/polyethylene component should not exceed the total thickness of the native patella [4,121].

Although the incidence of failure after TKA is low, approximately 20,000 to 30,000 knee replacements are revised yearly [109,122]. The most recent series analyzing failed TKAs find that more that 50% of revisions are undertaken in the first 2 years after primary surgery [122]. The mode of failure in more than half of these cases is the result of instability, misalignment, malposition, or failure of fixation [122]. Another larger, yet older, study demonstrates that most failures occur within 5 years after primary surgery, and the most common reasons for revision were infection and instability [123].

In the largest study to date, the overall infection rate after TKA is less than 2% [122–124]. When infection occurs, it almost always is deep and tends to occur in the early postoperative period [122–124]. Although elevations of the erythrocyte sedimentation rate and C-reactive protein serum levels are not entirely specific for infection, levels greater than 30 mm/h and 20 mg/L, respectively, at least are suggestive of infection and warrant further investigation [1,4,125,126].

Various scintigraphic examinations and techniques (labeled WBC scans, conventional bone scans, and sulfur colloid imaging) are studied for their abilities to determine the presence or absence of infection after TKA, with wide-ranging results

[127–129]. Because it is normal to see increased activity for up to 1 year after surgery, traditional technetium Tc 99m–methylene diphosphonate scanning is unreliable in distinguishing septic from aseptic loosening. It may, however, be of great value when the examination is negative or demonstrates decreasing activity in serial examinations [4,127]. Although routine WBC imaging is of limited use in this clinical setting, the addition of 24-hour delayed imaging increases positive predictive value modestly, and the addition of technetium sulfur colloid imaging increases overall accuracy dramatically to greater than 90% [128,129].

Some investigators report promising results using F18- fluorodeoxyglucose positron emission tomography (PET) in the evaluation of suspected infection after TKA, with intermediate uptake suggestive of loosening and intense uptake suggestive of infection [130,131]. Other investigators conclude that the usefulness of PET in this clinical setting is similar to that of bone scanning, and the most valuable information likely lies in a negative examination [132].

Despite the real or theoretic usefulness of these examinations (radiographs, scintigraphy, PET, and so forth) it is the authors' experience that most orthopedic surgeons ultimately desire imaging-guided joint aspiration before embarking on a revision arthroplasty. Accuracy of aspiration greater than 90% is reported in the literature; however, it is essential that all antibiotic therapy be discontinued for a minimum of 4 weeks before aspiration [4,133–136]. When revision arthroplasty is undertaken in the setting of infection, it is critical that any subclinical, low-grade infection be identified and eradicated before the second stage re-implantation procedure. This also should be performed via aspiration under the same stringent criteria [4,124–126].

The thickness of the polyethylene component is determined by the tension needed to balance the knee ligaments but should, at the outset, be at least 8 mm [4,137]. Polyethylene wear is a multifactorial phenomenon influenced by patient weight and activity level, polyethylene thickness, alignment, chemical and biomechanical properties of the polyethylene, and the relationship between the polyethylene component and the metal surfaces of the tibial and femoral components [137]. Polyethylene wear traditionally has been evaluated radiographically by measuring the distance from the femoral condyles to the tibial baseplate on standard AP and lateral projections and by evaluating change during the course of serial examinations. More recently, ultrasound is shown highly accurate in the evaluation of polyethylene thickness when correlated with radiographic findings [138]. Joint space loss and associated instability caused by polyethylene wear is problematic alone. Shed polyethylene debris and ensuing histiocytic response can complicate matters further.

Loosening is a well-understood and well-documented complication of TKA. Conventional radiographs may be adequate in determining the presence of this complication, particularly when studied during the course of serial examinations. Although cemented and noncemented tibial components normally may develop a thin (<1 mm) rim of lucency around the keel and baseplate, an enlarging rim of lucency or one measuring greater than 2 mm is abnormal and consistent with loosening [139]. The distinction, however, between aseptic loosening (ie, that resulting from mechanical wear) and septic loosening cannot reliably be made radiographically.

Although any of the components (metal, polyethylene, and cement) of a total knee system may invoke osteolysis, the most common cause is polyethylene wear [140]. Radiographically, osteolysis is seen as a new or expanding, typically well-defined area of lucency adjacent to the femoral or tibial (or less likely patellar) components [Fig. 16] [141]. In the past, extensive beam-hardening artifacts on CT and metallic susceptibility artifacts on MR imaging rendered these modalities almost useless in evaluation of patients who had pain after TKA. Current scanner technology and imaging software, however, have brought these previously limited modalities to the fore in the evaluation of patients who are post knee arthroplasty. In the largest study to date using MR imaging for the evaluation of post-TKA pain, therapeutically relevant information was obtained in a large number of cases [142].

With current surgical techniques and components, the incidence of patellar or extensor mechanism complications after TKA is less than 5% [143]. Potential complications include patellar maltracking, "overstuffing," patellar "clunk," patellar fracture, component loosening, and failure of the

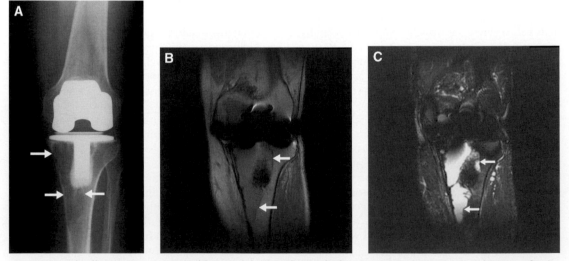

Fig. 16. Particle disease in a 73-year-old man. (A) AP radiograph demonstrates postoperative changes of prior TKA and well-defined periprosthetic lucency in the proximal tibia (arrows). Coronal FSE proton density (3000/23) (B) and FSE inversion recovery (C) MR images of the same patient demonstrate intermediate and increased signal, respectively (arrows). Findings are consistent with particle disease, which was confirmed at the time of revision arthroplasty. Osteolysis also is present along the periprosthetic distal femur.

extensor tendon mechanism [4]. Patellar maltracking may occur with as little as 4° of internal rotation of the femoral component and is assessed best on axial CT images. Greater than 4° of internal rotation may lead to patellar subluxation, dislocation, or failure of the patellar component [4,144]. Whether or not the tibial component should be aligned with the femoral/condylar component or placed intentionally in internal rotation is a controversial issue yet to be resolved in the orthopedic literature [145–147]. Patellar overstuffing occurs when the aggregate thickness of the remaining patella and the patellar component is greater than the thickness of the native patella or when the condylar component is too large [4]. This is problematic because of the increased stress placed on the lateral retinaculum, which ultimately may lead to maltracking or subluxation of the patella [148]. Over-resection of the native patella also is problematic because of the increased stress placed on the remaining patellar bone, which may result in anterior knee pain (when less than 15 mm) or fracture (when less than 11 mm) [143,148].

Patellar fracture after TKA is a rare complication, occurring in 0.68% of patients in a large series of more than 12,000 patients [Fig. 17] [149]. A more recent, smaller series in the radiology literature reports an incidence of 1.14% [150]. Causes of patellar fracture after TKA include osteonecrosis of the patella, over-resection of the native patella, and component misalignment or maltracking, both of which may place undue stress on the remaining patella [4]. Loosening of the patellar component is a complication encountered less frequently with current surgical techniques that use a cemented

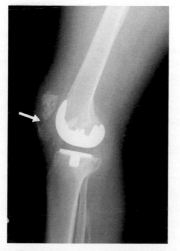

Fig. 17. Patellar fracture in a 62-year-old man after knee arthroplasty. Lateral radiograph demonstrates a fracture deformity through the lower pole of the left patella (*arrow*).

polyethylene component as opposed to the cemented or uncemented metallic components used in the past [4]. Failure of the extensor mechanism is a rare complication after TKA, occurring in less than 3% of cases, but typically necessitates surgical repair. MR imaging and sonography may be beneficial in detection of this complication and interrogation of the extensor tendons.

Akin to the cylops lesion after ACL tear/reconstruction, a small reactive fibrous nodule also may develop on the deep surface of the distal quadriceps tendon immediately above the superior pole of the patella after TKA. When a patient extends the knee, this nodule may snap out of the intercondylar notch where it can become trapped during flexion, manifesting as an audible clunk. This condition is referred to as the patellar clunk syndrome [4,151]. There are no reports in the literature regarding MR imaging evaluation of this complication; however, the appearance of a small echogenic mass with internal vascularity is described with sonography [152].

Although periprosthetic fractures may occur intra- or postoperatively, the latter most likely comes to radiologists' attention. The incidence of supracondylar femoral fractures above TKAs is reported to be between 0.3% and 2.5% [153]. These fractures may occur anytime between the immediate postoperative period and more than a decade after surgery, with a mean of 2 to 4 years [154]. Risk factors include osteopenia, femoral notching, and poor flexion [154]. Osteolysis and component loosening also are described as risk factors for the development of fracture [4]. Periprosthetic fractures of the proximal tibia or the condylar or tibial metal components are even rarer, occurring in only 0.2% of TKAs [155]. Complications unique to unicompartmental prostheses include the development of degenerative arthritis in the opposite compartment and stress fractures of the proximal tibia below the component [4].

After TKA or any operation requiring immobilization, patients are at risk for developing deep venous thrombosis. Although the diagnosis may be suggested by clinical examination (positive Homans' sign, palpable cord, and so forth) or by serologic criteria (eg, elevated D-dimer), most patients ultimately are referred for imaging evaluation. In this clinical setting, ultrasound is the preferred method of evaluation. Fat embolism syndrome after elective TKA is another rare, but potential, complication that may be associated with respiratory or neurologic compromise [156].

Some patients may be found, despite a thorough clinical and imaging evaluation, to have idiopathic pain after TKA [4]. The literature suggests that little is gained by re-operating on these patients, and these patients may be worse after a repeat operation [157–159]. It is possible that current advances in TKA surgical technique, including the use of pre- and intraoperative CT guidance and robotic assistance, will, as suggested by recent reports, advance the care of these patients further and decrease the number of patients who have recurrent or persistent pain [160–163].

Articular cartilage

Chondral injuries typically occur in association with other soft tissue injuries (eg, cruciate ligaments or menisci). These injuries rarely occur in isolation, particularly in skeletally immature patients [164,165]. Unlike many tissues in the human body, hyaline articular cartilage lacks the ability to repair itself and precocious osteoarthritis often is the outcome.

Although articular cartilage is not directly visualized radiographically, its absence or attrition may be implied by joint space narrowing. Sonography of articular cartilage is described [166]. By and large, however, the current imaging test of choice for evaluation of articular cartilage is MR imaging in the native and postoperative states. As summarized by Polster and colleagues, the articular cartilage is evident on all MR imaging pulse sequences, but two pulse sequences in particular (3-D SPGR T1 and fast spin-echo [FSE] with or without fat saturation) are shown by several investigators to have the highest sensitivity, specificity, and accuracy [2,167–172]. On 3-D SPGR and FSE images, normal articular cartilage is seen as a smooth band of hyperintense or intermediate signal intensity tissue, respectively, paralleling the underlying cortex [2]. Pathologic conditions manifest on 3-D SPGR images as areas of loss of the normal high signal intensity, and high signal intensity fluid outlines defects on FSE images [2]. Although FSE images offer the advantage of being less prone to the metallic susceptibility artifacts compared with gradient-echo images, the operative techniques to repair articular cartilage in common use today do not routinely involve placement of metallic components or devices. Although the signal intensity of the articular cartilage may vary depending on the sequence used, the morphology should not. Normal articular cartilage should have a smooth, congruent surface on any MR imaging pulse sequences. Any deviation from this appearance should be considered abnormal.

Chondral pathology includes chondromalacia, fibrillation, fissuring or flaps, chondral fractures, thinning, and displaced/loose intra-articular fragments [2]. Subchondral bone marrow edema implies a full thickness defect of the articular carti-

lage, even when not directly evident [2]. As in the detection of native cartilage lesions, MR imaging is the preferred imaging modality for monitoring postoperative changes. Recent advances in MR imaging technology and software allow real-time color mapping of the integrity of the articular cartilage [173–175].

Three main surgical options are available to repair chondral injuries: (1) bone marrow stimulation techniques, (2) osteochondral transplantation, and (3) chondrocyte transplantation. Unfortunately, none currently offers long-term solutions. Bone marrow stimulation techniques include abrasion arthroplasty and subchondral drilling, but the procedure performed most frequently is microfracture [Fig. 18] [2]. The basic tenet of all three techniques is that exposed subchondral bone marrow allows marrow elements, most importantly pluripotential stem cells and blood, to fill the chondral defect. Fibrin clot then serves as scaffolding on which chondrocytes and fibroblasts lay down a new layer of fibrocartilage [176–178].

The ideal postoperative appearance is complete filling of the defect with fibrocartilage that is congruent with the native, uninjured hyaline cartilage and restoration of a smooth, well-defined surface spanning the articular surface. In the initial postoperative period, the fibrocartilage may be thinner than the adjacent native cartilage; however, this discrepancy should resolve by 2 years [179]. In the immediate postoperative period, it is common, if not expected, to see bone marrow edema at the site of microfracture. This is a finding that usually resolves with time but usually is of no clinical significance in the rare instances in which it persists [179].

Although a difference in signal intensity between the fibrocartilage and native articular (or hyaline) cartilage is expected [176], any deviation from a smooth, congruent articular surface should be considered abnormal [2,176]. Abnormal postoperative MR imaging findings include incomplete filling of the gap, persistently thinned or progressively thinning repair tissue, gaps between the repair tissue and native cartilage, chondral flaps, or chondral irregularity [179].

Because of the poor biomechanical characteristics of reparative fibrocartilage, several techniques have been developed that provide hyaline or hyaline-like cartilage in the articular defect. These techniques include periosteal and perichondral grafts, autologous chondrocyte transplantation, morselized autologous osteochondral mixture, osteochondral allografts, biomaterials, and autologous osteochondral transplantation [180].

Osteochondral autograft transplantation surgery (eg, mosaicplasty) consists of harvesting cylindric

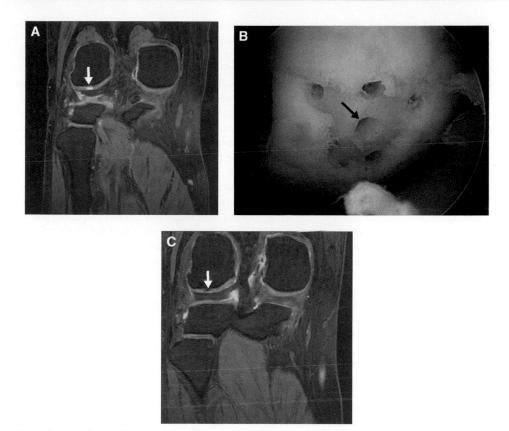

Fig. 18. Microfracture for cartilage defect. (*A*) Coronal fat-saturated T2-weighted MR images demonstrate a focal, near full-thickness chondral defect of the right lateral femoral condylar articular surface (*arrow*). (*B*) Arthroscopic image obtained during microfracture procedure demonstrates puncture defects (*arrow*) within the lateral femoral condyle. (*C*) After chondroplasty, this defect is filled with low-signal intensity fibrocartilage (*arrow*).

plugs of articular cartilage and underlying bone from less critical non–weight-bearing areas and transplanting them into size-matched surgical defects or holes in the region of the chondral defect [180]. Hangody and colleagues recently authored an illustrated review of this surgical technique to which readers are referred [181]. As described in their review and as suggested by others, it usually is necessary to use multiple plugs to fill the defect adequately [7]. The small crevices between the hyaline cartilage of the cylindrical plugs typically fill in with blood and fibrin clot with eventual deposition of fibrocartilage. As the articular cartilage often is thinner at the donor sites than at the defect, a small cortical offset of the transplanted plugs must be accepted to achieve the desired result of a smooth, congruent articular surface [Fig. 19] [182].

The articular hyaline cartilage of the transplanted osteochondral unit should, given its identical histologic makeup, demonstrate T2-weighted signal intensity similar to normal articular cartilage adjacent to the repaired defect but may, in some cases,

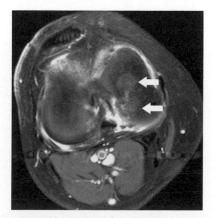

Fig. 19. Mosaicplasty. Axial fat-saturated T2-weighted MR image performed after mosaicplasty demonstrates mildly increased T2 signal intensity within the articular cartilage of the transplanted osteochondral plugs (*arrows*).

demonstrate areas of increased T2-weighted signal intensity [182]. Small areas of heterogeneous signal intensity are expected in the tiny crevices between the plugs, reflecting the presence of fibrocartilage at these locations. Increased T2-weighted signal intensity and enhancement are seen postoperatively at donor and recipient sites, believed to represent the fibrovascular tissue present in the healing response leading to graft incorporation [183]. This signal change usually resolves by 12 months but may be mildly persistent for up to 2 years [2]. Donor site defects demonstrate low T1- and high T2-weighted signal intensity postoperatively but eventually return to normal marrow signal [2].

As with the bone marrow stimulation techniques, the optimal postoperative appearance is that of a smooth, congruent articular surface. Any deviation from such an appearance is undesirable and typically leads to a poor outcome. Potential complications include incongruent positioning with protrusion or recession of the graft margins, graft subsidence, or failure of graft incorporation [2]. Complications at the donor site are unusual.

Autologous chondrocyte implantation (ACI) is another attempt to repair cartilage defects with generation of normal articular or hyaline cartilage. The technique involves multiple stages and includes harvesting a small population of chondrocytes, which are grown ex vivo in culture to a larger colony for a period of 4 to 6 weeks. Once the chondrocyte colony matures, an open procedure is performed. The defect is débrided to the level of cortical bone. A periosteal roof (typically harvested from the ipsilateral tibia or femur) is constructed, sewn in place, and sealed peripherally with glue; chondrocytes subsequently are injected into the covered defect [2,184–187]. Several stages of maturation are described, which include the proliferative phase (first 6 weeks), the transition phase (7–12 weeks), and the remodeling and maturation phase (up to 9–18 months), beyond which the repair tissue should be as firm as native cartilage [2,187,188].

MR imaging appearances after ACI depend on temporal relation to surgery [189]. Initially, the transplanted tissue appears heterogeneous and may remain so as the graft undergoes maturation and remodeling, but the graft should be nearly indiscernible from adjacent native cartilage at full maturation [Fig. 20] [2]. Again, the final postoperative result should be a smooth, uninterrupted, and congruent surface; any deviation from such should be considered abnormal. Postoperative complications unique to ACI include fibrous overgrowth of the periosteal cover, intra-articular adhesions, and, more ominously, delamination of the repair tissue [2,186]. Overgrowth of the periosteal roof may be smooth or may be protuberant and drape over the adjacent normal cartilage (edge overlap); this occurs in up to one quarter of all patients after ACI, with resultant mechanical symptoms [2]. Periosteal hypertrophy is treated via arthroscopic debridement. Delamination represents failure of incorporation of the graft or traumatic dissociation from shearing forces [2]. This complication is reported to occur in approximately 5% of patients, typically within the first 6 months [185,186]. At MR imaging, this is manifest as linear T2 hyperintensity, similar to fluid, between the graft and underlying bone [2]. Complete and large partial delamination injuries typically necessitate repeat ACI, whereas small, partial delamination injuries may be salvaged by microfracture [185]. As with the other repair techniques, it is common to see edema in the underlying subchondral bone marrow that resolves slowly with time. Edema persisting longer than 1 year or increasing with time is abnormal and may herald a problem with the transplant [2].

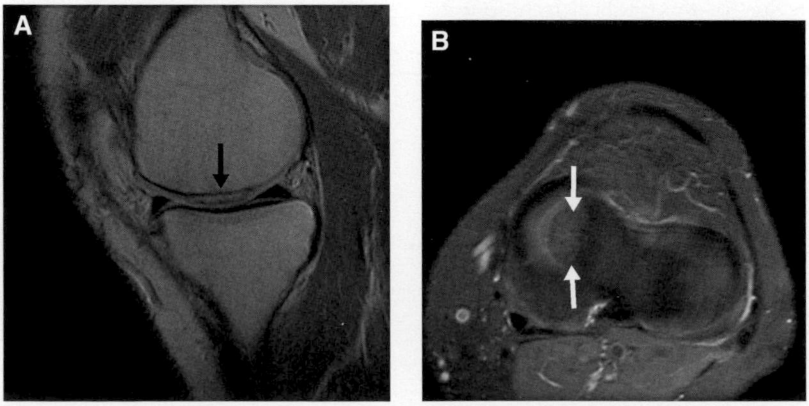

Fig. 20. ACI in a 35-year-old man. Proton density sagittal (A) and proton-density fat-suppressed axial (B) MR images obtained 12 months after ACI. The chondrocyte implant is relatively isointense to the native hyaline cartilage (arrows). Notice the thin rim of hypointensity, which represents the periosteal sleeve.

In summary, as evidenced by the large number of recent contributions to the radiology and orthopedic literature, imaging of the postoperative knee is a fascinating and rapidly evolving field. Knowledge of the expected postoperative appearance and potential complications of the procedures that are performed commonly will allow for improved communication between radiologists and orthopedists and, ideally, will lead to improved patient care.

References

[1] Toms AP, White LM, Marshall TJ, et al. Imaging the postoperative meniscus. Eur J Radiol 2005; 54:189–98.

[2] Polster J, Recht M, Magee T, et al. Postoperative MR evaluation of chondral repair in the knee. MR arthrography of postoperative knee: for which patients is it useful? Magnetic resonance imaging of autologous chondrocyte implantation. Eur J Radiol 2005;54:206–13.

[3] Motamedi K, Seeger LL, Hame SL. Imaging of postoperative knee extensor mechanism. Eur J Radiol 2005;54:199–205.

[4] Miller TT. Imaging of knee arthroplasty. Eur J Radiol 2005;54:164–77.

[5] Ilaslan H, Sundaram M, Miniaci A. Imaging evaluation of the postoperative knee ligaments. Eur J Radiol 2005;54:178–88.

[6] McCauley TR. MR imaging evaluation of the postoperative knee. Radiology 2005;234:53–61.

[7] White LM, Kramer J, Recht MP. MR imaging evaluation of the postoperative knee: ligaments, menisci, and articular cartilage. Skeletal Radiol 2005;34:431–52.

[8] Sanders TG. MR imaging of postoperative ligaments of the knee. Semin Musculoskel Radiol 2002;6:19–33.

[9] Vande Berg BC, Lecouvet FE, Poilvache P, et al. Spiral CT arthrography of the postoperative knee. Semin Musculoskel Radiol 2002;6:47–55.

[10] Recht MP, Parker RD, Irizarry JM. Second time around: evaluating the postoperative anterior cruciate ligament. Magn Reson Imaging Clin North Am 2000;8:285–97.

[11] Kannus P, Jarvinen M. Conservatively treated tears of the anterior cruciate ligament. Long-term results. J Bone Joint Surg [Am] 1987;69-A: 1007–12.

[12] McDaniel Jr WJ, Dameron Jr TB. The untreated anterior cruciate ligament rupture. Clin Orthop 1983;172:158–63.

[13] Finsterbush A, Frankl U, Matan Y, et al. Secondary damage to the knee after isolated injury of the anterior cruciate ligament. Am J Sports Med 1990;18:475–9.

[14] Messner K, Maletius W. Eighteen- to twenty-five-year follow-up after acute partial anterior cruciate ligament rupture. Am J Sports Med 1999;27:455–9.

[15] Ferretti A, Conteduca F, De Carli A, et al. Osteo-arthritis of the knee after ACL reconstruction. Int Orthop 1991;15:367–71.

[16] Guidoin MF, Marois Y, Bejui J, et al. Analysis of retrieved polymer fiber based replacements for the ACL. Biomaterials 2000;21:2461–74.

[17] Giron F, Aglietti P, Cuomo P, et al. Anterior cruciate ligament reconstruction with double-looped semitendinosus and gracilis tendon graft directly fixed to cortical bone: 5-year results. Knee Surg Sports Traumatol Arthrosc 2005;13:81–91.

[18] Williams 3rd RJ, Hyman J, Petrigliano F, et al. Anterior cruciate ligament reconstruction with a four-strand hamstring tendon autograft. Surgical technique. J Bone Joint Surg [Am] 2005; 87-A(Suppl 1[Pt 1]):51–66.

[19] Herrington L, Wrapson C, Matthews M, et al. Anterior Cruciate Ligament reconstruction, hamstring versus bone-patella tendon-bone grafts: a systematic literature review of outcome from surgery. Knee 2005;12:41–50.

[20] Laxdal G, Kartus J, Hansson L, et al. A prospective randomized comparison of bone-patellar tendon-bone and hamstring grafts for anterior cruciate ligament reconstruction. Arthroscopy 2005;21:34–42.

[21] Brand Jr J, Weiler A, Caborn DN, et al. Graft fixation in cruciate ligament reconstruction. Am J Sports Med 2000;28:761–74.

[22] Wagner M, Kaab MJ, Schallock J, et al. Hamstring tendon versus patellar tendon anterior cruciate ligament reconstruction using biodegradable interference fit fixation: a prospective matched-group analysis. Am J Sports Med 2005; 33:1327–36.

[23] Tomczak RJ, Hehl G, Mergo PJ, et al. Tunnel placement in anterior cruciate ligament reconstruction: MRI analysis as an important factor in the radiological report. Skeletal Radiol 1997;26: 409–13.

[24] Howell SM, Berns GS, Farley TE. Unimpinged and impinged anterior cruciate ligament grafts: MR signal intensity measurements. Radiology 1991;179:639–43.

[25] Howell SM, Clark JA. Tibial tunnel placement in anterior cruciate ligament reconstructions and graft impingement. Clin Orthop 1992;283: 187–95.

[26] Howell SM, Clark JA, Farley TE. A rationale for predicting anterior cruciate graft impingement by the intercondylar roof. A magnetic resonance imaging study. Am J Sports Med 1991;19:276–82.

[27] Howell SM, Gittins ME, Gottlieb JE, et al. The relationship between the angle of the tibial tunnel in the coronal plane and loss of flexion and anterior laxity after anterior cruciate ligament reconstruction. Am J Sports Med 2001;29: 567–74.

[28] Bents RT, Jones RC, May DA, et al. Intercondylar notch encroachment following anterior cruciate ligament reconstruction: a prospective study. Am J Knee Surg 1998;11:81–8.

[29] Mann TA, Black KP, Zanotti DJ, et al. The

natural history of the intercondylar notch after notchplasty. Am J Sports Med 1999;27:181–8.

[30] Vergis A, Gillquist J. Graft failure in intra-articular anterior cruciate ligament reconstructions: a review of the literature. Arthroscopy 1995;11:312–21.

[31] Schatz JA, Potter HG, Rodeo SA, et al. MR imaging of anterior cruciate ligament reconstruction. AJR Am J Roentgenol 1997;169:223–8.

[32] Rak KM, Gillogly SD, Schaefer RA, et al. Anterior cruciate ligament reconstruction: evaluation with MR imaging. Radiology 1991;178:553–6.

[33] Yamato M, Yamagishi T. MRI of patellar tendon anterior cruciate ligament autografts. J Comput Assist Tomogr 1992;16:604–7.

[34] Stockle U, Hoffmann R, Schwedke J, et al. Anterior cruciate ligament reconstruction: the diagnostic value of MRI. Int Orthop 1998;22:288–92.

[35] McCauley TR, Elfar A, Moore A, et al. MR arthrography of anterior cruciate ligament reconstruction grafts. AJR Am J Roentgenol 2003;181:1217–23.

[36] Trattnig S, Rand T, Czerny C, et al. Magnetic resonance imaging of the postoperative knee. Top Magn Reson Imaging 1999;10:221–36.

[37] Vogl TJ, Schmitt J, Lubrich J, et al. Reconstructed anterior cruciate ligaments using patellar tendon ligament grafts: diagnostic value of contrast-enhanced MRI in a 2-year follow-up regimen. Eur Radiol 2001;11:1450–6.

[38] Papakonstantinou O, Chung CB, Chanchairujira K, et al. Complications of anterior cruciate ligament reconstruction: MR imaging. Eur Radiol 2003;13:1106–17.

[39] Howell SM, Taylor MA. Failure of reconstruction of the anterior cruciate ligament due to impingement by the intercondylar roof. J Bone Joint Surg [Am] 1993;75-A:1044–55.

[40] Olson PN, Rud P, Griffiths HJ. Cyclops lesion. Orthopedics 1995;18:1041.

[41] Recht MP, Piraino DW, Cohen MA, et al. Localized anterior arthrofibrosis (cyclops lesion) after reconstruction of the anterior cruciate ligament: MR imaging findings. AJR Am J Roentgenol 1995;165:383–5.

[42] Bradley DM, Bergman AG, Dillingham MF. MR imaging of cyclops lesions. AJR Am J Roentgenol 2000;174:719–26.

[43] Carson EW, Anisko EM, Restrepo C, et al. Revision anterior cruciate ligament reconstruction: etiology of failures and clinical results. J Knee Surg 2004;17:127–32.

[44] Brettler D, Soudry M. Tibial bone plug resorption with extra-articular cyst: a rare complication of anterior cruciate ligament reconstruction. Arthroscopy 1995;11:478–81.

[45] Rokkanen PU, Bostman O, Hirvensalo E, et al. Bioabsorbable fixation in orthopaedic surgery and traumatology. Biomaterials 2000;21:2607–13.

[46] Bostman O, Partio E, Hirvensalo E, et al. Foreign-body reactions to polyglycolide screws. Observations in 24/216 malleolar fracture cases. Acta Orthop Scand 1992;63:173–6.

[47] Yanmis I, Tunay S, Oguz E, et al. Dropping of an EndoButton into the knee joint 2 years after anterior cruciate ligament repair using proximal fixation methods. Arthroscopy 2004;20:641–3.

[48] Yoshiya S, Andrish JT, Manley MT, et al. Graft tension in anterior cruciate ligament reconstruction. An in vivo study in dogs. Am J Sports Med 1987;15:464–70.

[49] Muneta T, Sekiya I, Yagishita K, et al. Two-bundle reconstruction of the anterior cruciate ligament using semitendinosus tendon with endobuttons: operative technique and preliminary results. Arthroscopy 1999;15:618–24.

[50] May DA, Snearly WN, Bents R, et al. MR imaging findings in anterior cruciate ligament reconstruction: evaluation of notchplasty. AJR Am J Roentgenol 1997;169:217–22.

[51] Cosgarea AJ, DeHaven KE, Lovelock JE. The surgical treatment of arthrofibrosis of the knee. Am J Sports Med 1994;22:184–91.

[52] Cerullo G, Puddu G, Gianni E, et al. Anterior cruciate ligament patellar tendon reconstruction: it is probably better to leave the tendon defect open! Knee Surg Sports Traumatol Arthrosc 1995;3:14–7.

[53] Bernicker JP, Haddad JL, Lintner DM, et al. Patellar tendon defect during the first year after anterior cruciate ligament reconstruction: appearance on serial magnetic resonance imaging. Arthroscopy 1998;14:804–9.

[54] Marumoto JM, Mitsunaga MM, Richardson AB, et al. Late patellar tendon ruptures after removal of the central third for anterior cruciate ligament reconstruction. A report of two cases. Am J Sports Med 1996;24:698–701.

[55] DuMontier TA, Metcalf MH, Simonian PT, et al. Patella fracture after anterior cruciate ligament reconstruction with the patellar tendon: a comparison between different shaped bone block excisions. Am J Knee Surg 2001;14:9–15.

[56] Wilk KE, Andrews JR, Clancy WG. Quadriceps muscular strength after removal of the central third patellar tendon for contralateral anterior cruciate ligament reconstruction surgery: a case study. J Orthop Sports Phys Ther 1993;18:692–7.

[57] Fahey M, Indelicato PA. Bone tunnel enlargement after anterior cruciate ligament replacement. Am J Sports Med 1994;22:410–4.

[58] Wilson TC, Kantaras A, Atay A, et al. Tunnel enlargement after anterior cruciate ligament surgery. Am J Sports Med 2004;32:543–9.

[59] Jagodzinski M, Foerstemann T, Mall G, et al. Analysis of forces of ACL reconstructions at the tunnel entrance: is tunnel enlargement a biomechanical problem? J Biomech 2005;38:23–31.

[60] Parolie JM, Bergfeld JA. Long-term results of

nonoperative treatment of isolated posterior cruciate ligament injuries in the athlete. Am J Sports Med 1986;14:35–8.

[61] Keller PM, Shelbourne KD, McCarroll JR, et al. Nonoperatively treated isolated posterior cruciate ligament injuries. Am J Sports Med 1993;21: 132–6.

[62] Shelbourne KD, Rubinstein Jr RA, VanMeter CD, et al. Correlation of remaining patellar tendon width with quadriceps strength after autogenous bone-patellar tendon-bone anterior cruciate ligament reconstruction. Am J Sports Med 1994;22:774–7 [discussion: 777–8].

[63] Sherman PM, Sanders TG, Morrison WB, et al. MR imaging of the posterior cruciate ligament graft: initial experience in 15 patients with clinical correlation. Radiology 2001;221:191–8.

[64] Mariani PP, Adriani E, esca G. Arthroscopic-assisted posterior cruciate ligament reconstruction using patellar tendon autograft: a technique for graft passage. Arthroscopy 1996;12:510–2.

[65] Mariani PP, Adriani E, Santori N, et al. Arthroscopic posterior cruciate ligament reconstruction with bone-tendon-bone patellar graft. Knee Surg Sports Traumatol Arthrosc 1997;5:239–44.

[66] Mariani PP, Adriani E, Bellelli A, et al. Magnetic resonance imaging of tunnel placement in posterior cruciate ligament reconstruction. Arthroscopy 1999;15:733–40.

[67] Wind Jr WM, Bergfeld JA, Parker RD. Evaluation and treatment of posterior cruciate ligament injuries: revisited. Am J Sports Med 2004;32: 1765–75.

[68] Buess E, Imhoff AB, Hodler J. Knee evaluation in two systems and magnetic resonance imaging after operative treatment of posterior cruciate ligament injuries. Arch Orthop Trauma Surg 1996;115:307–12.

[69] Buzzi R, Aglietti P, Vena LM, et al. Lateral collateral ligament reconstruction using a semi-tendinosus graft. Knee Surgery Sports Traumatol Arthrosc 2004;12:36–42.

[70] Kim SJ, Park IS, Cheon YM, et al. New technique for chronic posterolateral instability of the knee: posterolateral reconstruction using the tibialis posterior tendon allograft. Arthroscopy 2004;20(Suppl 2):195–200.

[71] Latimer HA, Tibone JE, ElAttrache NS, et al. Reconstruction of the lateral collateral ligament of the knee with patellar tendon allograft. Report of a new technique in combined ligament injuries. Am J Sports Med 1998;26:656–62.

[72] Chen CH, Chen WJ, Shih CH. Lateral collateral ligament reconstruction using quadriceps tendon-patellar bone autograft with bioscrew fixation. Arthroscopy 2001;17:551–4.

[73] Borden PS, Kantaras AT, Caborn DN. Medial collateral ligament reconstruction with allograft using a double-bundle technique. Arthroscopy 2002;18:E19.

[74] Hede A, Larsen E, Sandberg H. The long term outcome of open total and partial meniscec-tomy related to the quantity and site of the meniscus removed. Int Orthop 1992;16:122–5.

[75] Newman AP, Daniels AU, Burks RT. Principles and ision making in meniscal surgery. Arthroscopy 1993;9:33–51.

[76] Johnson MJ, Lucas GL, Dusek JK, et al. Isolated arthroscopic meniscal repair: a long-term outcome study (more than 10 years). Am J Sports Med 1999;27:44–9.

[77] Roos H, Lauren M, Adalberth T, et al. Knee osteoarthritis after meniscectomy: prevalence of radiographic changes after twenty-one years, compared with matched controls. Arthritis Rheum 1998;41:687–93.

[78] Roos EM, Ostenberg A, Roos H, et al. Long-term outcome of meniscectomy: symptoms, function, and performance tests in patients with or without radiographic osteoarthritis compared to matched controls. Osteoarthritis Cartilage 2001;9:316–24.

[79] Rodkey WG. Basic biology of the meniscus and response to injury. Instr Course Lect 2000;49: 189–93.

[80] Magee T, Shapiro M, Williams D. Prevalence of meniscal radial tears of the knee revealed by MRI after surgery. AJR Am J Roentgenol 2004; 182:931–6.

[81] DeHaven KE. Meniscus repair. Am J Sports Med 1999;27:242–50.

[82] Baratz ME, Rehak DC, Fu FH, et al. Peripheral tears of the meniscus. The effect of open versus arthroscopic repair on intraarticular contact stresses in the human knee. Am J Sports Med 1988;16:1–6.

[83] Weiss CB, Lundberg M, Hamberg P, et al. Nonoperative treatment of meniscal tears. J Bone Joint Surg [Am] 1989;71-A:811–22.

[84] Arnoczky SP, Warren RF. Microvasculature of the human meniscus. Am J Sports Med 1982;10: 90–5.

[85] Eggli S, Wegmuller H, Kosina J, et al. Long-term results of arthroscopic meniscal repair. An analysis of isolated tears. Am J Sports Med 1995;23:715–20.

[86] Applegate GR, Flannigan BD, Tolin BS, et al. MR diagnosis of recurrent tears in the knee: value of intraarticular contrast material. AJR Am J Roentgenol 1993;161:821–5.

[87] Kent RH, Pope CF, Lynch JK, et al. Magnetic resonance imaging of the surgically repaired meniscus: six-month follow-up. Magn Reson Imaging 1991;9:335–41.

[88] Deutsch AL, Mink JH, Fox JM, et al. Peripheral meniscal tears: MR findings after conservative treatment or arthroscopic repair. Radiology 1990;176:485–8.

[89] Farley TE, Howell SM, Love KF, et al. Meniscal tears: MR and arthrographic findings after arthroscopic repair. Radiology 1991;180:517–22.

[90] Lim PS, Schweitzer ME, Bhatia M, et al. Repeat tear of postoperative meniscus: Potential MR imaging signs. Radiology 1999;210:183–8.

[91] Smith DK, Totty WG. The knee after partial meniscectomy: MR imaging features. Radiology 1990;176:141–4.

[92] Fischer SP, Fox JM, Del Pizzo W, et al. Accuracy of diagnoses from magnetic resonance imaging of the knee. A multi-center analysis of one thousand and fourteen patients. J Bone Joint Surg [Am] 1991;73-A:2–10.

[93] Crues 3rd JV, Mink J, Levy TL, et al. Meniscal tears of the knee: accuracy of MR imaging. Radiology 1987;164:445–8.

[94] Schulte-Altedorneburg G, Gebhard M, Wohlgemuth WA, et al. MR arthrography: pharmacology, efficacy and safety in clinical trials. Skeletal Radiol 2003;32:1–12.

[95] Magee T, Shapiro M, Rodriguez J, et al. MR arthrography of postoperative knee: for which patients is it useful? Radiology 2003;229:59–63.

[96] Sciulli RL, Boutin RD, Brown RR, et al. Evaluation of the postoperative meniscus at the knee: a study comparing conventional arthrography, conventional MR imaging, MR arthrography with iodinated contrast material, and MR arthrography with gadolinium-based contrast material. Skeletal Radiol 1999;28:508–14.

[97] White LM, Schweitzer ME, Weishaupt D, et al. Diagnosis of recurrent meniscal tears: prospective evaluation of conventional MR imaging, indirect MR arthrography, and direct MR arthrography. Radiology 2002;222:421–9.

[98] Vives MJ, Homesley D, Ciccotti MG, et al. Evaluation of recurring meniscal tears with gadolinium-enhanced magnetic resonance imaging: a randomized, prospective study. Am J Sports Med 2003;31:868–73.

[99] Mutschler C, Vande Berg BC, Lecouvet FE, et al. Postoperative meniscus: assessment at dual-detector row spiral CT arthrography of the knee. Radiology 2003;228:635–41.

[100] Milachowski KA, Weismeier K, Wirth CJ. Homologous meniscus transplantation. Experimental and clinical results. Int Orthop 1989;13:1–11.

[101] Wirth CJ, Peters G, Milachowski KA, et al. Long-term results of meniscal allograft transplantation. Am J Sports Med 2002;30:174–81.

[102] Noyes FR, Barber-Westin SD, Rankin M. Meniscal transplantation in symptomatic patients less than fifty years old. J Bone Joint Surg [Am] 2005;87-A(Suppl 1[Pt 2]):149–65.

[103] Potter HG, Rodeo SA, Wickiewicz TL, et al. MR imaging of meniscal allografts: correlation with clinical and arthroscopic outcomes. Radiology 1996;198:509–14.

[104] Verstraete KL, Verdonk R, Lootens T, et al. Current status and imaging of allograft meniscal transplantation. Eur J Radiol 1997;26:16–22.

[105] van Arkel ER, Goei R, de Ploeg I, et al. Meniscal allografts: evaluation with magnetic resonance imaging and correlation with arthroscopy. Arthroscopy 2000;16:517–21.

[106] Faletti C, Robba T, de Petro P. Postmeniscectomy osteonecrosis. Arthroscopy 2002;18:91–4.

[107] DeFalco RA, Ricci AR, Balduini FC. Osteonecrosis of the knee after arthroscopic meniscectomy and chondroplasty: a case report and literature review. Am J Sports Med 2003;31:1013–6.

[108] Brander VA, Stulberg SD, Adams AD, et al. Predicting total knee replacement pain: a prospective, observational study. Clin Orthop 2003;416:27–36.

[109] Mahomed NN, Barrett J, Katz JN, et al. Epidemiology of total knee replacement in the United States medicare population. J Bone Joint Surg [Am] 2005;87-A:1222–8.

[110] Vince KG, Cyran LT. Unicompartmental knee arthroplasty: new indications, more complications? J Arthroplasty 2004;19(4, Suppl 1):9–16.

[111] Deshmukh RV, Scott RD. Unicompartmental knee arthroplasty: long-term results. Clin Orthop 2001;392:272–8.

[112] Burnett RS, Haydon CM, Rorabeck CH, et al. Patella resurfacing versus nonresurfacing in total knee arthroplasty: results of a randomized controlled clinical trial at a minimum of 10 years' followup. Clin Orthop 2004;428:12–25.

[113] Bourne RB, Burnett RS. The consequences of not resurfacing the patella. Clin Orthop 2004;428:166–9.

[114] Pakos EE, Ntzani EE, Trikalinos TA. Patellar resurfacing in total knee arthroplasty. A meta-analysis. J Bone Joint Surg [Am] 2005;87-A:1438–45.

[115] Weir DJ, Moran CG, Pinder IM. Kinematic condylar total knee arthroplasty. 14-year survivorship analysis of 208 consecutive cases. J Bone Joint Surg [Br] 1996;78-B:907–11.

[116] Keating EM, Meding JB, Faris PM, et al. Long-term followup of nonmodular total knee replacements. Clin Orthop 2002;404:34–9.

[117] Gill GS, Joshi AB, Mills DM. Total condylar knee arthroplasty. 16- to 21-year results. Clin Orthop 1999;367:210–5.

[118] Buechel Sr FF. Long-term followup after mobile-bearing total knee replacement. Clin Orthop 2002;404:40–50.

[119] Manaster BJ. Total knee arthroplasty: postoperative radiologic findings. AJR Am J Roentgenol 1995;165:899–904.

[120] Lesh ML, Schneider DJ, Deol G, et al. The consequences of anterior femoral notching in total knee arthroplasty. A biomechanical study. J Bone Joint Surg [Am] 2000;82-A:1096–101.

[121] Allen AM, Ward WG, Pope Jr TL. Imaging of the total knee arthroplasty. Radiol Clin North Am 1995;33:289–303.

[122] Sharkey PF, Hozack WJ, Rothman RH, et al. Insall Award paper. Why are total knee arthroplasties failing today? Clin Orthop 2002;404:7–13.

[123] Fehring TK, Odum S, Griffin WL, et al. Early failures in total knee arthroplasty. Clin Orthop 2001;392:315–8.

[124] Peersman G, Laskin R, Davis J, et al. Infection

in total knee replacement: a retrospective review of 6489 total knee replacements. Clin Orthop 2001;392:15–23.

[125] Barrack RL, Jennings RW, Wolfe MW, et al. The Coventry Award. The value of preoperative aspiration before total knee revision. Clin Orthop 1997;345:8–16.

[126] Sanzen L, Carlsson AS. The diagnostic value of C-reactive protein in infected total hip arthroplasties. J Bone Joint Surg [Br] 1989;71-B:638–41.

[127] Smith SL, Wastie ML, Forster I. Radionuclide bone scintigraphy in the detection of significant complications after total knee joint replacement. Clin Radiol 2001;56:221–4.

[128] Larikka MJ, Ahonen AK, Junila JA, et al. Improved method for detecting knee replacement infections based on extended combined 99mTc-white blood cell/bone imaging. Nucl Med Commun 2001;22:1145–50.

[129] Joh TN, Mujtaba M, Chen AL, et al. Efficacy of combined technetium-99m sulfur colloid/indium-111 leukocyte scans to detect infected total hip and knee arthroplasties. J Arthroplasty 2001;16:753–8.

[130] Manthey N, Reinhard P, Moog F, et al. The use of [18 F]fluorodeoxyglucose positron emission tomography to differentiate between synovitis, loosening and infection of hip and knee prostheses. Nucl Med Commun 2002;23:645–53.

[131] Zhuang H, Duarte PS, Pourdehnad M, et al. The promising role of 18F-FDG PET in detecting infected lower limb prosthesis implants. J Nucl Med 2001;42:44–8.

[132] Love C, Pugliese PV, Afriyie MO, et al. 5. Utility of F-18 FDG Imaging for diagnosing the infected joint replacement. Clin Positron Imaging 2000;3:159.

[133] Duff GP, Lachiewicz PF, Kelley SS. Aspiration of the knee joint before revision arthroplasty. Clin Orthop 1996;331:132–9.

[134] Mont MA, Waldman BJ, Hungerford DS. Evaluation of preoperative cultures before second-stage reimplantation of a total knee prosthesis complicated by infection. A comparison-group study. J Bone Joint Surg [Am] 2000;82-A:1552–7.

[135] Lonner JH, Beck Jr TD, Rees H, et al. Results of two-stage revision of the infected total knee arthroplasty. Am J Knee Surg 2001;14:65–7.

[136] Lonner JH, Siliski JM, Della Valle C, et al. Role of knee aspiration after resection of the infected total knee arthroplasty. Am J Orthop 2001;30:305–9.

[137] Goodman S, Lidgren L. Polyethylene wear in knee arthroplasty. A review. Acta Orthop Scand 1992;63:358–64.

[138] Sofka CM, Adler RS, Laskin R. Sonography of polyethylene liners used in total knee arthroplasty. AJR Am J Roentgenol 2003;180:1437–41.

[139] Fehring TK, McAvoy G. Fluoroscopic evaluation of the painful total knee arthroplasty. Clinic Orthop 1996;331:226–33.

[140] Naudie DD, Rorabeck CH. Sources of osteolysis around total knee arthroplasty: wear of the bearing surface. Instr Course Lect 2004;53:251–9.

[141] Berry DJ. Recognizing and identifying osteolysis around total knee arthroplasty. Instr Course Lect 2004;53:261–4.

[142] Sofka CM, Potter HG, Figgie M, et al. Magnetic resonance imaging of total knee arthroplasty. Clinic Orthop Rel Res 2003;406:129–35.

[143] Holt GE, Dennis DA. The role of patellar resurfacing in total knee arthroplasty. Clin Orthop 2003;416:76–83.

[144] Berger RA, Crossett LS, Jacobs JJ, et al. Malrotation causing patellofemoral complications after total knee arthroplasty. Clinic Orthop 1998;356:144–53.

[145] Barrack RL, Schrader T, Bertot AJ, et al. Component rotation and anterior knee pain after total knee arthroplasty. Clin Orthop 2001;392:46–55.

[146] Berger RA, Rubash HE. Rotational instability and malrotation after total nee arthroplasty. Orthop Clin North Am 2001;32:639–47.

[147] Uehara K, Kadoya Y, Kobayashi A, et al. Bone anatomy and rotational alignment in total knee arthroplasty Clin Orthop 2002;402:196–201.

[148] Malo M, Vince KG. The unstable patella after total knee arthroplasty: etiology, prevention, and management. J Am Acad Orthop Surg 2003;11:364–71.

[149] Ortiguera CJ, Berry DJ. Patellar fracture after total knee arthroplasty. J Bone Joint Surg [Am] 2002;84-A:532 40.

[150] Chun KA, Ohashi K, Bennett DL, et al. Patellar fractures after total knee replacement. AJR Am J Roentgenol 2005;185:655–60.

[151] Maloney WJ, Schmidt R, Sculco TP. Femoral component design and patellar clunk syndrome. Clin Orthop 2003;410:199–202.

[152] Okamoto T, Futani H, Atsui K, et al. Sonographic appearance of fibrous nodules in patellar clunk syndrome: a case report. J Orthop Sci 2002;7:590–3.

[153] Hayakawa K, Nakagawa K, Ando K, et al. Ender nailing for supracondylar fracture of the femur after total knee arthroplasty: five case reports. J Arthroplasty 2003;18:946–52.

[154] Tharani R, Nakasone C, Vince KG. Periprosthetic fractures after total knee arthroplasty. J Arthroplasty 2005;20(Suppl 2):27–32.

[155] Huang CH, Ma HM, Lee YM, et al. Long-term results of low contact stress mobile-bearing total knee replacements. Clin Orthop 2003;416:265–70.

[156] Jenkins K, Chung F, Wennberg R, et al. Fat embolism syndrome and elective knee arthroplasty. Can J Anesth 2002;49:19–24.

[157] Mont MA, Serna FK, Krackow KA, et al. Exploration of radiographically normal total knee replacements for unexplained pain. Clin Orthop 1996;331:216–20.

[158] Jacobs MA, Hungerford DS, Krackow KA, et al.

Revision total knee arthroplasty for aic failure. Clin Orthop 1988;226:78–85.

[159] Dennis DA. Evaluation of painful total knee arthroplasty. J Arthroplasty 2004;19(4, Suppl 1): 35–40.

[160] King J, Theis C, Achenbach T, et al. Robotic total knee arthroplasty: the accuracy of CT-based component placement. Acta Orthop Scand 2004; 75:573–9.

[161] Victor J, Hoste D. Image-based computer-assisted total knee arthroplasty leads to lower variability in coronal alignment. Clin Orthop 2004;428:131–9.

[162] Chauhan SK, Clark GW, Lloyd S, et al. Computer-assisted total knee replacement. A controlled cadaver study using a multi-parameter quantitative CT assessment of alignment (the Perth CT Protocol). J Bone Joint Surg [Br] 2004; 86-B:818–23.

[163] Swank ML. Computer-assisted surgery in total knee arthroplasty:recent advances. Surg Technol Int 2004;12:209–13.

[164] Terry GC, Flandry F, Van Manen JW, et al. Isolated chondral fractures of the knee. Clin Orthop 1988;234:170–7.

[165] Oeppen RS, Connolly SA, Bencardino JT, et al. Acute injury of the articular cartilage and subchondral bone: a common but unrecognized lesion in the immature knee. AJR Am J Roentgenol 2004;182:111–7.

[166] Saied A, Lier P. High-resolution ultrasonography for analysis of age- and disease-related cartilage changes. Methods Mol Med 2004;101: 249–65.

[167] Recht MP, Kramer J, Celis S, et al. Abnormalities of articular cartilage in the knee: analysis of available MR techniques. Radiology 1993;187: 473–8.

[168] Disler DG, McCauley TR, Kelman CG, et al. Fat-suppressed three-dimensional spoiled gradient-echo MR imaging of hyaline cartilage defects in the knee: comparison with standard MR imaging and arthroscopy. AJR Am J Roentgenol 1996;167:127–32.

[169] Disler DG, McCauley TR, Wirth CR, et al. Detection of knee hyaline cartilage defects using fat-suppressed three-dimensional spoiled gradient-echo MR imaging: comparison with standard MR imaging and correlation with arthroscopy. AJR Am J Roentgenol 1995;165: 377–82.

[170] Recht MP, Piraino DW, Paletta GA, et al. Accuracy of fat-suppressed three-dimensional spoiled gradient-echo FLASH MR imaging in the detection of patellofemoral articular cartilage abnormalities. Radiology 1996;198:209–12.

[171] Potter HG, Linklater JM, Allen AA, et al. Magnetic resonance imaging of articular cartilage in the knee. An evaluation with use of fast-spin-echo imaging. J Bone Joint Surg [Am] 1998; 80-A:1276–84.

[172] Bredella MA, Tirman PF, Peterfy CG, et al. Accuracy of T2-weighted fast spin-echo MR imaging with fat saturation in detecting cartilage defects in the knee: comparison with arthroscopy in 130 patients. AJR Am J Roentgenol 1999;172:1073–80.

[173] Mosher TJ, Smith H, Dardzinski BJ, et al. MR imaging and T2 mapping of femoral cartilage: in vivo determination of the magic angle effect. AJR Am J Roentgenol 2001;177:665–9.

[174] Mosher TJ, Smith HE, Collins C, et al. Change in knee cartilage T2 at MR imaging after running: a feasibility study. Radiology 2005;234: 245–9.

[175] Van Breuseghem I, Bosmans HT, Elst LV, et al. T2 mapping of human femorotibial cartilage with turbo mixed MR imaging at 1.5 T: feasibility. Radiology 2004;233:609–14.

[176] Nehrer S, Spector M, Minas T. Histologic analysis of tissue after failed cartilage repair procedures. Clin Orthop 1999;365:149–62.

[177] Steadman JR, Rodkey WG, Briggs KK. Microfracture to treat full-thickness chondral defects: surgical technique, rehabilitation, and outcomes. J Knee Surg 2002;15:170–6.

[178] Steadman JR, Rodkey WG, Rodrigo JJ. Microfracture: surgical technique and rehabilitation to treat chondral defects. Clin Orthop 2001; 391(Suppl):S362–9.

[179] Alparslan L, Winalski CS, Boutin RD, et al. Postoperative magnetic resonance imaging of articular cartilage repair. Semin Musculoskelet Radiol 2001;5:345–63.

[180] Hangody L, Feczko P, Bartha L, et al. Mosaicplasty for the treatment of articular defects of the knee and ankle. Clin Orthop 2001;39(Suppl): S328–36.

[181] Hangody L, Rathonyi GK, Duska Z, et al. Autologous osteochondral mosaicplasty. Surgical technique. J Bone Joint Surg [Am] 2004; 86-A(Suppl 1):65–72.

[182] Motamedi K, Seeger LL, Hame SL, et al. Imaging of postoperative knee extensor mechanism. Imaging evaluation of the postoperative knee ligaments. Eur J Radiol 2005;54:199–205.

[183] Sanders TG, Mentzer KD, Miller MD, et al. Autogenous osteochondral "plug" transfer for the treatment of focal chondral defects: postoperative MR appearance with clinical correlation. Skeletal Radiol 2001;30:570–8.

[184] Peterson L, Brittberg M, Kiviranta I, et al. Autologous chondrocyte transplantation. Biomechanics and long-term durability. Am J Sports Med 2002;30:2–12.

[185] Minas T, Peterson L. Advanced techniques in autologous chondrocyte transplantation. Clin Sports Med 1999;18:13–44.

[186] Peterson L, Minas T, Brittberg M, et al. Two- to 9-year outcome after autologous chondrocyte transplantation of the knee. Clin Orthop 2000; 374:212–34.

[187] Minas T, Chiu R. Autologous chondrocyte implantation. Am J Knee Surg 2000;13:41–50.

[188] Richardson JB, Caterson B, Evans EH, et al. Repair of human articular cartilage after implantation of autologous chondrocytes. J Bone Joint Surg [Br] 1999;81-A:1064–8.

[189] Recht M, White LM, Winalski CS, et al. MR imaging of cartilage repair procedures. Skeletal Radiol 2003;32:185–200.

[190] Yoshida S, Recht MP. Postoperative evaluation of the knee. Radiol Clin North Am 2002;40: 1133–46.

RADIOLOGIC
CLINICS
OF NORTH AMERICA

Radiol Clin N Am 44 (2006) 391–406

Postoperative Imaging of the Ankle and Foot

Diane Bergin, MD*, William B. Morrison, MD

Familiarity with foot and ankle surgery and the expected imaging features of these procedures is essential to avoid diagnostic pitfalls and ensure accurate assessment of the osseous and soft tissue structures. To evaluate post-therapeutic patients critically, it is important to understand the primary clinical diagnosis, surgical treatment undergone by patients, the interval since surgery, and patients' current clinical symptoms. The most common imaging modality for evaluating the postoperative ankle and foot is radiography. This article discusses when MR imaging may be useful for evaluating the postoperative foot and ankle. It illustrates the expected and abnormal postsurgical MR imaging appearance of the foot and ankle. This discussion provides the background, practical information, and graphic examples necessary to better enable radiologists to approach the clinicoradiologic MR evaluation of the postoperative foot and ankle.

MR imaging

Because of its excellent contrast discrimination capabilities, MR imaging has an important role in evaluating osseous and soft tissues after surgery and in diagnosing postoperative complications. MR imaging, however, is susceptible to image degradation and loss of detail secondary to the presence of surgical hardware. As there is in-depth discussion of techniques to improve imaging artifact elsewhere in this issue, a few items are mentioned here. Susceptibility artifacts may be reduced by using spin-echo or fast spin-echo sequences, a higher bandwidth, and a short echo time. Use of frequency-selective fat suppression and gradient-echo imaging should be limited where possible. Short tau inversion recovery (STIR) is a fluid-sensitive sequence that is less sensitive to magnetic field inhomogeneities and is a useful alternative to frequency-selective

Department of Radiology, Thomas Jefferson University Hospital, 111 South 10th Street, Philadelphia, PA 19107, USA
* Corresponding author.
E-mail address: diane.bergin@jefferson.edu (D.Bergin).

Table 1: Imaging the postoperative ankle with 1.5-Tesla magnet

Sequence	FOV	Matrix	TR	TE	TI	Bandwidth	ETL
Long-axis PD non–fat saturated	16	256 × 192	400–800	3000	–	16	–
Long-axis T2 FSE fat saturated	16	256 × 256	>2000	20–40	–	16	16
Sagittal T1	16	256 × 192	400–800	Min	–	16	–
Sagittal STIR	14	256 × 192	>2000	20–40	150	16	16
Short-axis 2-D FMSPGR pre- and postcontrast	12	256 × 192	200	4	–	16	–
Sagittal and coronal 2-D FMSPGR	12	256 × 192	200	4	–	16	–

Abbreviations: ETL, echo train length; FMSPGR, fast multiplanar spoiled gradient-recalled echo; FOV, field of view; Min, minimum; PD, proton density; TI, time to inversion.

fat suppression. Images acquired after intravenous gadolinium also can improve soft tissue and osseous detail in the presence of metallic susceptibility artifact. If artifact from hardware is extensive and no other imaging modality is feasible, radiologists may consider MR imaging with a lower-strength magnet of 0.7 Tesla or less.

Standard MR imaging examinations of the ankle and forefoot with 1.5 Tesla are performed at the authors' institution using a transmit-receive coil (Signa, General Electric Medical Systems, Milwaukee, Wisconsin). The patients are imaged in a supine position, feet first, with the ankle in mild plantar flexion. The standard MR imaging protocol for the ankle includes sagittal T1-weighted (repetition time [TR]/echo time [TE] 500–750/22) and STIR (4000–5000/21) sequences, axial T1-weighted (700–900/22) and fast spin-echo, fat-suppressed, T2-weighted (TR range/TE 3000–3500/80–90) sequences, and coronal T1-weighted images (TR range/TE 600–700/22). The field of view is 12 to 14 cm. The slice thickness is 3 to 4 mm with an interslice gap of 1 mm and matrix of 512 × 254 for T1-weighted and 256 × 256 for T2-weighted images. The standard MR imaging protocol for the forefoot includes sagittal T1-weighted (TR/TE 500–750/22) and STIR (4000–5000/21) sequences, coronal T1-weighted (700–900/22) and fast spin-echo, fat-suppressed, T2-weighted (TR/TE 3000–3500/80–90) sequences, and axial T1-weighted images (TR/TE 600–700/22). Similar field of view, slice thickness, and matrix parameters are used as for the ankle. The authors advocate separate im-

aging of the forefoot and ankle. Intravenous contrast is administered when there is questionable infection, suspicion of devitalized tissues, or recurrent mass [Tables 1 and 2]. Intravenous gadolinium also provides greater soft tissue detail when there is metallic susceptibility artifact in the anatomic area of interest.

Tendon surgery

General principles

The goal of repairing tendon injuries is to produce a union of adequate tensile strength and to restore gliding function as quickly as possible. Atraumatic technique is a primary principle of operative repair [1,2]. Many tendon disruptions are treated by surgical repair and postoperative immobilization. An ideal suturing technique apposes the tendon ends and allows early passive motion without gapping [2]. The initial healing of an injured tendon progresses over a 4-week period. During the first week, the severed ends are joined loosely by granulation tissue. Paratenon vascularity increases during the second week and some passive electrically stimulated contractions may be noted. Collagen fibrils align longitudinally during the third week to provide moderate strength, and in the fourth week, edema recedes and guarded limited motion is possible [2,3].

After 8 weeks of interrupted function, muscle becomes contracted. To restore length to a ruptured tendon, a graft may be necessary [2]. Autografts are

Table 2: Imaging the postoperative forefoot with 1.5-Tesla magnet

Sequence	FOV	Matrix	TR	TE	TI	Bandwidth	ETL
Short-axis T1 non–fat saturated	12	256 × 192	400–800	Min	—	16	—
Short-axis T2 FSE fat saturated	12	256 × 256	>2000	20–40	—	16	16
Sagittal STIR	14	256 × 192	>2000	20–40	150	16	16
Short-axis 2-D FMSPGR pre- and postcontrast	12	256 × 192	200	4	—	16	—
Sagittal and coronal 2-D FMSPGR	12	256 × 192	200	4	—	16	—

Abbreviations: FMSPGR, fast multiplanar spoiled gradient-recalled echo; FSE, fast spin echo.

preferred and when a longer graft is required, the plantaris, peroneus tertius, or peroneus brevis tendons may be used [3]. To bridge shorter gaps, portions of the extensor digitorum longus, extensor digitorum brevis, or extensor hallucis brevis (EHB) are suitable [3,4]. The diameter of the graft should match that of the recipient tendon as closely as possible. This helps reduce adhesions, because each end of the injured tendon extends fibrotic projections into the intervening repair site [4].

The transplanted tendon initially acts as a fibrotic strut and eventually becomes incorporated into the recipient tendon. Although the graft initially is avascular, by the end of the first week, new capillaries penetrate and begin the revitalization process. By 10 to 12 weeks, the graft should appear identical to the recipient tendon [2].

Achilles tendon

The Achilles tendon is the largest and strongest tendon in the human body. Ruptures of the Achilles tendon occur most commonly spontaneously in healthy, young, active individuals 20 to 50 years old who have no antecedent history of calf or heel pain [5]. The majority of Achilles tendon injuries are a result of indirect trauma [5].

Considerable controversy exists regarding whether or not operative or nonoperative treatment should be used in the treatment of acute ruptures [6]. There are no universal criteria for operative intervention or for the best method of repair. Cetti and colleagues randomized a group of 111 patients to operative and nonoperative treatment and report that the surgically managed group had a significantly higher rate of resuming sports activities, a lesser degree of calf atrophy, and fewer complaints 1 year after rupture [7].

Surgical repair of Achilles tendon

Many surgical techniques for repair of Achilles tendon rupture are described [8–10]. Surgery for Achilles tendon can be open or percutaneous [9,11]. Bradley and Tibone compare percutaneous and open techniques and recommend percutaneous repair in recreational athletes [11]. Because of the slightly higher rerupture rate with the percutaneous technique, open surgical technique is recommended for higher caliber athletes [11].

Partial tears or tears with a less than 3-cm tendon gap may be repaired by end-to-end anastomosis [10,11]. This typically is performed using a suture technique [12]. Patients who have a gap of 3 to 6 cm at the rupture site are repaired using an autogenous tendon graft flap. If the gap is greater than 6 cm, a free tendon graft or synthetic graft is advocated.

A neglected Achilles tendon rupture of greater than 4 weeks' duration requires surgical repair [8]. Techniques for repair include fascia lata reconstruction, woven gastrocnemius aponeurosis, gastrocnemius flap turndown procedures, plantaris tendon reinforcements, peroneus brevis tendon transfer, carbon synthetic graft, flexor digitorum longus (FDL) or flexor hallucis longus (FHL) transfer, and Marlex mesh graft [8]. These techniques also may be used in the acute setting to augment a tenuous tendon repair.

Achilles tendon lengthening is used in patients who have diabetes with recurrent plantar ulcers resulting from increased plantar pressure and a component of gastrocnemius equines [13]. Achilles lengthening increases dorsiflexion of the foot and decreases forefoot shear forces. Percutaneous tendon lengthening involves making several small stab wounds through the skin and nicking the tendon so that it tears partially and then is longer. Z-plasty lengthening involves pen incision behind the ankle [14,15]. A Z-cut is made in the tendon. The tendon is stretched apart and then sutured [15,16].

Surgical treatment of Haglund's syndrome may be necessary when symptoms are resistant to standard conservative therapy. Surgery entails resection of the bony protuberance of the posterior calcaneus (Haglund's deformity) and the associated adventitial bursa. This frequently also entails surgical débridement and repair of Achilles tear with reattachment of the tendon to the modified posterior calcaneus.

Imaging Achilles tendon repair

MR imaging may be used to evaluate the Achilles tendon after surgical repair [17,18]. At the authors' institution, standard MR imaging is performed using a 1.5-Tesla magnet with a modified field of view (16 cm or more on sagittal images) to incorporate the musculotendinous junction. The imaging appearance of the Achilles tendon after surgical repair is well described in the literature [18,19]. The principal MR imaging findings 1 year after surgery include generalized thickening and moderate heterogeneity [18,19]. Edema, tendon defects, and peritendinous reactions are seen in a minority of patients [18,19]. Augmentation of Achilles tendon repair by adjacent tendons, such as the FHL, peroneus brevis, or FDL, may be recognized by change in orientation of their standard course [Fig. 1]. Tendon defects or retears are recognized as localized, intratendinous, high-signal intensity areas on proton density- and T2-weighted sequences [Fig. 2] [20,21].

Other postoperative complications that may be imaged by MR imaging include postoperative infection that can lead to failed surgical repair or ossifi-

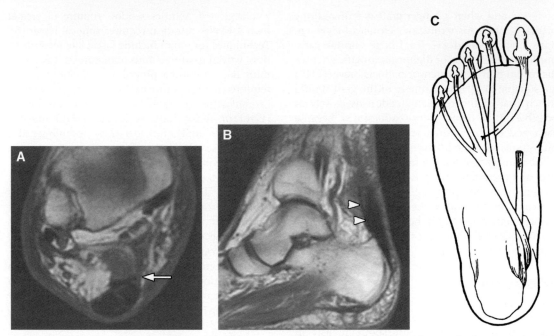

Fig. 1. FHL tendon transfer for Achilles tendon rupture. A 40-year-old man who has history of prior surgical repair of Achilles rupture presents with ankle pain unrelated to Achilles tendon. (*A*) Axial T1-weighted (TR/TE 600/22) image shows surgically repaired Achilles tendon with FHL tendon transfer (*arrow*). (*B*) Sagittal T1-weighted (TR/TE 500/22) image shows surgically repaired Achilles tendon with FHL tendon transfer (*arrowheads*). (*C*) Illustration demonstrates expected postsurgical appearance of the distal FHL in the forefoot after FHL transfer with tenodesis of the distal tendon to the FDL.

cation of the tendon leading to dysfunction [22]. Ossification of the Achilles tendon is depicted as areas of marrow signal within the tendon [Fig. 3]. It predisposes patients to retear. When infection is suspected, postcontrast images should be obtained to delineate potential abscess [Fig. 4], evaluate for osteomyelitis, and aid in differentiation between postoperative scarring and devitalized tissue. A suture granuloma of surgically repaired Achilles

tendon may have hyperintense signal around it like a foreign body and can be indistinguishable from a small abscess.

Tibialis anterior

Rupture of the tibialis anterior (TA) tendon is unusual but occurs typically 1 to 2 cm proximal to its insertion onto the medial cuneiform and base of the first metatarsal. This area has a more tenuous

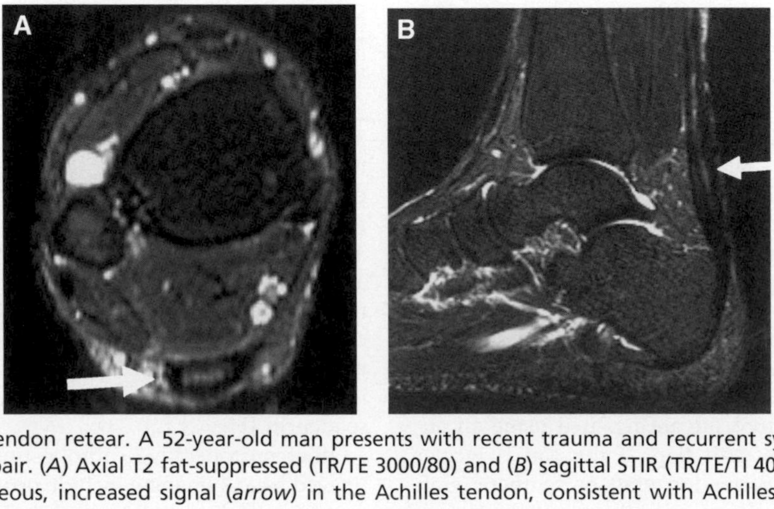

Fig. 2. Achilles tendon retear. A 52-year-old man presents with recent trauma and recurrent symptoms 2 years after surgical repair. (*A*) Axial T2 fat-suppressed (TR/TE 3000/80) and (*B*) sagittal STIR (TR/TE/TI 4000/30/15) images show heterogeneous, increased signal (*arrow*) in the Achilles tendon, consistent with Achilles tendon interstitial retear.

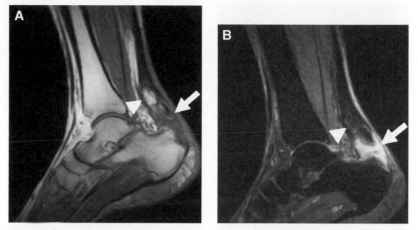

Fig. 3. Achilles tendon retear. A 60-year-old man who has history of remote Achilles tendon repair presents with recurrent symptoms. (*A*) Sagittal T1-weighted (TR/TE 500/22) and (*B*) STIR (TR/TE/TI 4000/30/15) images show diffuse thickening of the Achilles tendon. Signal intensity in the distal tendon equivalent to bone marrow (*arrowhead*) is the result of heterotopic ossification. There is discontinuity of the tendon at its insertion consistent with retear (*arrow*).

blood supply. Disruption also may occur between the superior and inferior retinacula or at the musculotendinous junction. Conservative management is recommended for avulsion injuries with osseous fragments where displacement of tendon edges is less than 5 mm. Elderly or less active patients may be treated with a below-the-knee non–weight-bearing cast with the foot dorsiflexed and inverted for 4 to 6 weeks.

Surgical repair is advocated for most patients, especially those who are young and middle aged and present within 3 to 4 months after injury. If presentation is early, end-to-end repair with non-absorbable sutures usually is possible. If there is a significant tendon gap precluding direct primary repair, then a tendon graft may be used. These include a free tendon graft from the extensor digitorum longus, a sliding tendon lengthening using a split portion of the proximal intact TA tendon, and a free peroneus brevis tendon graft along with peritendinous tissue. Another technique involves suturing the proximal stump of the TA to the adjacent extensor hallucis longus (EHL) tendon, severing the extensor tendon at the metatarsophangeal joint, and rerouting the extensor to the site of the insertion of the TA tendon. The distal portion of the EHL tendon then is anastomosed to the second digit or sutured to the EHB tendon.

Extensor hallucis longus

Lacerations of the EHL tendon are the result of sharp objects having an impact on a relatively un-protected tendon on the dorsum of the foot. Ruptures are rare but may occur because of sudden plantar flexor force applied to extended hallux.

Some investigators advocate that formal repair of EHL tendon may not be necessary because extensor tendons tend to heal spontaneously [2,20,23]. If treated early, end-to-end anastomosis usually is feasible. In cases of extensive tendon retraction or delayed repair, a free tendon graft of the EHB may be used with end-to-end or side-to-side repair. When the EHL is avulsed completely from its attachment to the bases of the distal hallux, then operative attachment is preferred.

Extensor digitorum longus

Rupture of the extensor tendons supplying the lesser toes is rare. Disruption is the result of open wound or laceration. If the entire extensor tendon is ruptured or lacerated, then the leg may be immobilized for 4 weeks with a below-the-knee device or undergo open repair. When individual segments of terminal branches are compromised, then conservative management may be adequate.

Tibialis posterior

Direct laceration of the tibialis posterior tendon is uncommon. Disruption in association with open and closed fractures of the medial malleolus is reported [24,25]. Isolated traumatic rupture is rare. A watershed region of low vascularity is located behind the medial malleolus and is a common site for rupture. Most commonly, rupture is the result of chronic stress or acute stress on an already degenerated tendon. Rupture of the posterior tibialis tendon (PTT) is one of the most common causes for acquired flatfoot in the adult.

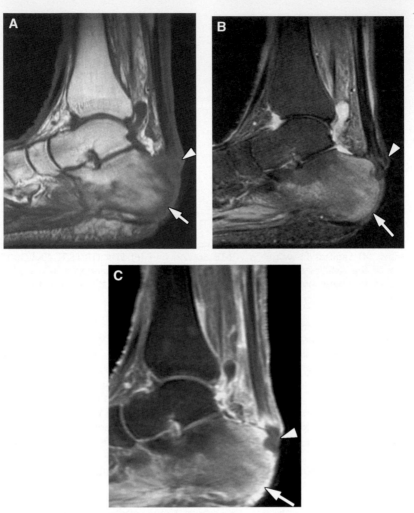

Fig. 4. Achilles tendon tear and osteomyelitis after calcaneal osteotomy. A 50-year-old woman who has persistent ankle pain and weakness 3 weeks after surgery for Haglund's syndrome. (*A*) Sagittal T1 (TR/TE 500/22) and (*B*) STIR (TR/TE/TI 4000/30/15) images show heterogeneous signal (*arrowhead*) within the Achilles tendon insertion, consistent with tear. There is replacement of the normal posterior calcaneal T1-weighted fat signal and edema-like marrow signal on STIR image, consistent with osteomyelitis (*arrow*). (*C*) Sagittal fat-suppressed, gradient-echo image after intravenous gadolinium shows a peripheral enhancing fluid collection at the Achilles insertion (*arrowhead*) and surgical site, consistent with postoperative abscess and Achilles tendon tear. There is marked enhancement of marrow within the calcaneus, consistent with osteomyelitis (*arrow*).

A large number of surgical procedures is described to address tibialis posterior dysfunction [23]. Tibialis tendon tears or dysfunction can be repaired surgically by primary repair or grafting. Longitudinal splits can be sutured. When the defect is large, it may be reinforced with the FDL tendon or with a portion of the TA by means of the Cobb procedure. The FDL is traced distally and released completely. Some surgeons may attach the distal stump of the FDL to the FHL to preserve some function of long flexor to the lesser toes. This is of debatable benefit. The FDL and tibialis posterior are sutured together and enclosed in the latter's sheath. The FDL also may be attached to the navicular [Fig. 5]. In the Cobb procedure, the TA tendon is split longitudinally from its insertion to the ankle. The medial aspect is released at the ankle and sutured to the tibialis posterior while maintaining its insertion on the medial cuneiform.

In the case of pes planus deformity or navicular cuneiform fault, the procedures (described previously) may be combined with the Young suspension to enhance arch stability [1,16]. The Young suspension involves rerouting the TA tendon through the navicular in a keyhole slot while maintaining its insertion. This creates a powerful plantar

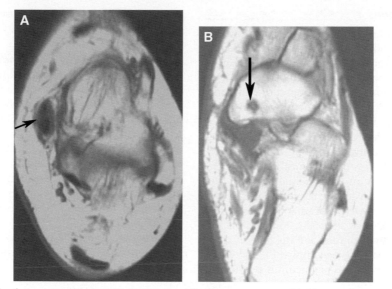

Fig. 5. FDL transfer for repair of torn PTT. A 52-year-old woman presents with history of surgery for pes planus. (*A*) Axial T1-weighted (TR/TE 600/22) image delineates the FDL tendon transfer, with the torn PTT and FDL tendons within the one sheath (*arrow*). (*B*) Axial T1-weighted image (TR/TE 600/22) demonstrates the suture site of the FDL onto the medial navicular (*arrow*).

navicular cuneiform ligament. Anastomosis of the peroneus brevis to the peroneus longus can be performed as an adjunctive procedure.

Long-term correction requires not only repair of the underlying tendon but also repair of the underling instability. Triple arthrodesis, once the mainstay of tibialis posterior tendon dysfunction repair, remains a viable procedure for the severely collapsed pes valgus deformity [1,26]. The principles of surgical reconstruction for adult-acquired flatfoot secondary to PTT dysfunction involve correcting hindfoot valgus deformity with lateral column lengthening procedure or medializing calcaneal osteotomy. Lengthening the lateral column can be accomplished by bone graft at the calcaneocuboid joint or distal calcaneus. Midfoot or forefoot varus may be corrected with medial column stabilization through arthrodesis of the first tarsometatarsal and naviculocuneiform joints. Medializing calcaneal osteotomy currently is the most favored surgical approach, but patients can complain of lateral foot pain. Osseous reconstruction may be combined with soft tissue procedures, such as FDL transfer for reconstruction of the PTT and correction of equinus deformity by gastrocnemius or tendo-Achilles lengthening [27].

Flexor hallucis longus

The FHL tendon is susceptible to injury along its entire course from the posterior ankle to its insertion. If laceration occurs, primary repair is recommended. Longitudinal tears in the midsubstance of the FHL may occur at the knot of Henry where the

tendon is under considerable strain, like a rope through a pulley. This may develop after acute or repetitive trauma as a result of hyperextension of the metatarsophalangeal joint. As conservative management usually fails to relieve symptoms, surgical repair is performed through medial exposure to the midfoot. Surgical repair involves release of the knot of Henry, excision of the interconnecting tendon, and repair of the longitudinal tendon split. Partial rupture of only the central fibers of the FHL is seen in classical ballet dancers and soccer players, resulting from erosion of the tendon by the sharp edge of the retinaculum. This problem may be addressed surgically through a posteromedial curvilinear excision.

When a disrupted tendon results in a forefoot FHL tear that cannot be apposed without excessive plantar flexion of the hallux, a Z-lengthening may be performed to avoid a flexion deformity of the hallux flexor tendon graft, or the fascia of the tensor fascia lata may be used. Complete rupture of the FHL within the tarsal canal may be treated with resection of the diseased tendon and tenodesis to the FDL tendon proximal and distal to the sheath [see Fig. 1C]. Tenodesis of the distal FHL to the flexor hallucis brevis tendon also can be used.

Flexor digitorum longus

Disruption of the FDL tendon typically is the result of laceration. Tears may be treated by primary repair or conservatively with immobilization therapy depending on the extent of rupture.

Peroneal tendons

Chronic longitudinal attrition or rupture of the peroneus brevis tendon may occur as single or multiple lesions. Chronic tears of the peroneus longus tendon seem to be the result of overuse without an inciting event with an insidious onset of symptoms [28]. Complete closed rupture of one or both peroneal tendons is rare. Disruption of these tendons usually is the result of strong tendinous contractions against an actively inverting foot or it may occur in association with a calcaneal fracture. Conservative treatment of longitudinal tears of the peroneal tendons may involve below-the-knee weight-bearing or a non–weight-bearing cast for 6 to 8 weeks. When treated surgically, longitudinal degenerative tears are débrided and the tendon tabularized using a small-gauge suture [3,29]. When present, excision of the os peroneum is performed. A sharp posterior osseous fibular ridge associated with peroneal brevis lesions also is resected. The posterior fibular groove may be deepened. The anterior portion of a longitudinally split peroneal brevis tendon may be excised. In the presence of ankle instability, the longitudinally split peroneal brevis may be divided proximally and used to perform ankle stabilization [3,29]. Surgical repair of completely disrupted tendons is recommended [30]. A plantaris graft may be used to bridge the gap between retracted tendon ends. In cases of complete disruption of the peroneal brevis tendon, the tear may be resected and the proximal tendon end attached to the peroneus longus tendon [3,29]. Complete rupture of the peroneal longus and brevis tendons may be treated by tenodesis to the cuboid or calcaneus, respectively, with reinforcement using a tendon graft [31]. Recurrent dislocation of the peroneal tendons may be treated surgically by peroneal groove deepening, tenoplasty, or bone block.

Imaging the postoperative tendon

As with the Achilles tendon, MR imaging is useful for assessing the postoperative appearance of the flexor and extensor tendons of the foot and ankle. Standard MR imaging protocol of the ankle or foot typically is sufficient, incorporating intravenous contrast when there is clinical concern for infection or to improve soft tissue detail when there is magnetic susceptibility. The repaired tendon is thickened in the immediate postsurgical period. Mechanism of surgical repair may be recognized by site of susceptibility artifact, tendon transfers, and osseous tunnels. Tendon defects or tears give rise to increased signal intensity on proton density and STIR sequences [Fig. 6].

Lateral ankle ligament reconstruction

Surgery for lateral ankle ligament injury is performed in patients who have repetitive inversion ankle sprains despite conservative therapy [32,33]. The MR appearance of the ankle after lateral ankle ligament reconstruction can be confusing because many surgical procedures are performed for ankle stabilization [32,33]. Familiarity with common surgical procedures and their imaging appearances is important to avoid misinterpreting postsurgical changes as tendon or bone disease and to ensure accurate assessment of the ligament reconstruction [Table 3].

Surgical options for reconstruction of the lateral ankle ligaments include direct lateral ligament repair, rerouting of the peroneus brevis tendon,

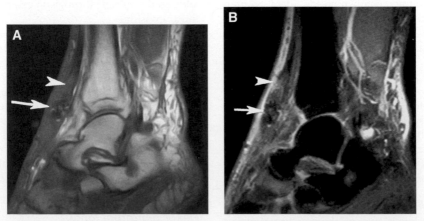

Fig. 6. Anterior tibial tendon retear. A 57-year-old man who has prior history of anterior tibialis tendon repair after ATT rupture. Patient presents with anterior ankle pain. (*A*) Sagittal T1-weighted (TR/TE 600/22) and (*B*) STIR (TR/TE/TI 4000/30/15) images demonstrate susceptibility artifact (*arrow*) anterior to the ankle at site of ATT repair. There is rerupture of the ATT, with discontinuity and tendinous retraction (*arrowhead*). Edema of the navicular is consistent with associated fracture.

Table 3: Common surgical procedures for repair of lateral ligaments and their pertinent MR imaging features

Surgical procedure		Fibular tunnel	Location of suture artifact	Peroneal brevis tendon	Additional findings
Brostrom	Direct lateral ligament repair	None	Lateral ligament	Normal	None
Modified Evans	Peroneal brevis tendon rerouting	Oblique vertical	Proximal PBT anastamosis	Proximal PBT anastamosis	None
Lee	Peroneal brevis tendon loop	Horizontal anteroposterior tunnel	Fibular tip PBT proximal anastamosis to PLT	Ligated proximally and anastomosed to PLT	PBT lateral to lateral malleolus
Watson-Jones	Peroneal brevis tendon loop	2 horizontal anteroposterior tunnel	Fibular tip PBT proximal anastamosis to PLT	Ligated proximally and anastomosed to PLT	PBT lateral to lateral malleolus
Chrisman-Snook	Peroneal brevis tendon splitting and rerouting	Horizontal AP fibular tunnel	Calcaneus talus base 5th metatarsal	Split longitudinally One component ligated to repair ligament	None

Abbreviations: PBT, peroneal brevis tendon; PLT, peroneal longus tendon.

peroneus brevis tendon loop, and peroneus brevis tendon splitting and rerouting [32,33]. Many surgical procedures designed to stabilize the lateral ankle are described [34–36]. One surgical treatment for lateral ankle ligament abnormality involves direct repair of the anterior talofibular ligament (ATFL) and possibly of the calcaneofibular ligament (CFL) if torn. In cases of severe ankle laxity, adjacent soft tissues, such as the peroneus brevis tendon, are used to augment ligament repair [3].

Direct lateral ligament repair

The Brostrom procedure involves direct repair of the ATFL; the two ends of the torn ligament are sutured together [37]. Direct repair of the CFL for ligament injury also may be performed [37]. The lateral extensor retinaculum, the lateral talocalcaneal ligament, or periosteal flaps may be used to reinforce the strength of the repair (modified Brostrom procedure) when subtalar laxity is present. Radiographs delineate suture anchors in the region of the ATFL. At MR imaging, there is corresponding artifact seen from the suture anchors where the ATFL attaches to the fibula. This may simulate the appearance of a fibular tunnel [37]. Careful evaluation of all sequences, in particular spin-echo sequences without fat suppression, should avoid this pitfall.

Peroneal brevis tendon rerouting

In peroneus brevis tendon rerouting (modified Evans procedure), the tendon is transected above the ankle, orientated in a distal-to-proximal direc-

tion through a surgically created tunnel in the fibula, and reattached at the initial site of transection [see Fig. 4]. This is a modification of the original procedure described by Evans, in which the transected peroneus brevis tendon was attached directly to the distal fibula to reconstruct the ATFL [38]. The oblique vertical fibular tunnel is an indicator on imaging that this procedure has been performed. The peroneus brevis tendon can be traced along its course from the fifth metatarsal through the fibular osseous tunnel on MR imaging. The tendon is sutured to the proximal peroneus at the site of transection superiorly. Artifact from the suture material is present at the reattachment site.

Peroneal brevis tendon loop

In a peroneus brevis tendon loop procedure (Lee procedure), the tendon is transected above the ankle and rerouted through the fibula [Fig. 7B]. The peroneus brevis tendon encircles the distal fibula and is sutured distally back on itself [24,26,32,33]. The fibular tunnel has an anteroposterior orientation on imaging. The peroneus brevis tendon courses proximally from the base of the fifth metatarsal, extends laterally to the lateral malleolus, and enters the fibular tunnel posteriorly. After exiting the anterior end of the tunnel, it is sutured to itself near the fibular tip. A periosteal flap of the adjacent fibula may be used to augment the reconstruction site. The proximal belly of the peroneus brevis muscle is sutured to the peroneus longus tendon. The Watson-Jones procedure [see Fig. 7C] is a modification of the peroneus brevis

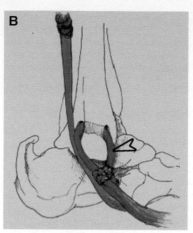

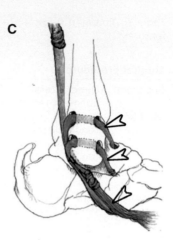

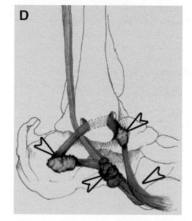

Fig. 7. Lateral ligament reconstructions. Diagrammatic illustrations of more commonly used surgical procedures for lateral ligament reconstruction using the peroneal brevis tendon. (*A*) Modified Evans procedure. Peroneal brevis tendon is transected proximally and passed through a fibular tunnel to be reanastomosed at the proximal transection site (*arrowheads*). (*B*) Lee procedure. Peroneal brevis tendon is transected proximally. The distal tendon is looped through a fibular tunnel and sutured to itself at the fibular tip (*arrowheads*). The proximal peroneus brevis is sutured to the peroneal longus. (*C*) Watson-Jones procedure. Peroneal brevis tendon is transected proximally. Distal tendon is looped trough two fibular tunnels and a tarsal tunnel and then sutured to itself at the fibular tip (*arrowheads*). (*D*) Chrisman-Snook procedure. Peroneal brevis tendon is split. Half the split tendon is passed beneath talar periosteum, fibular tunnel and calcaneal periosteum and then anastomosed to the peroneal longus and intact peroneal brevis beneath the lateral malleolus (*arrowheads*). (*From* Chien AJ, Jacobson JA, Jamadar DA, et al. Imaging appearance of lateral ligament reconstruction. RadioGraphics 2004;24: 999–1008; with permission.)

tendon loop with two fibular tunnels and a talar tunnel [24,26,32,33,39].

Peroneal brevis tendon split and rerouting

In peroneus brevis tendon split and rerouting (Chrisman-Snook procedure), the tendon is split longitudinally [see Fig. 7D]. Half of the tendon is separated proximally from the intact peroneus brevis muscle-tendon unit. From the base of the fifth metatarsal, it extends beneath the talus. It then passes through a fibular tunnel in an anteroposterior direction under the calcaneus [32,38]. The tendon is brought anteriorly and sutured on itself near the fifth metatarsal, with reconstruction of the ATFL

and CFL [33,40]. Maintaining integrity of the other half of the peroneus brevis tendon allows preservation of the dynamic function of the peroneus brevis muscle. The split peroneus brevis tendon can be recognized on MR imaging [Fig. 8]. The detached component can be traced proximally from the base of the fifth metatarsal to the talus, through the fibular tunnel to the calcaneus, and then back to itself, where artifact from suture material is evident.

Plantar fascia

Plantar fasciitis is a common source of pain in the foot, typically secondary to chronic microtrauma at

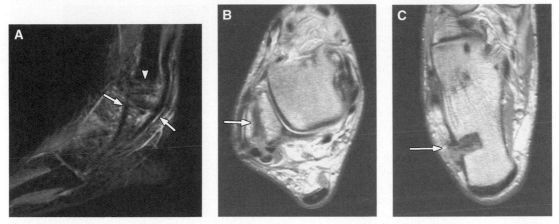

Fig. 8. Chrisman-Snook procedure. A 40-year-old man asymptomatic 2 years after lateral ligament reconstruction. (*A*) Sagittal STIR (TR/TE/TI 4000/30/15) image shows peroneal brevis splitting (*arrows*) and fibular tunnel (*arrowhead*). (*B*) Axial T1-weighted (TR/TE 500/22) image shows the fibular tunnel extending in an anteroposterior orientation (*arrow*). (*C*) Axial T1-weighted (TR/TE 500/22) image depicts the lateral calcaneal surgical defect (*arrow*) at anchor site for peroneal brevis tendon attachment.

the enthesis of the plantar fascia. Treatment generally is conservative. Patients whose symptoms are resistant to conservative therapy generally respond to operative management, with reported success rates of 90% to 95%. The plantar fascia may be released through an open incision or endoscopically. With the latter technique, a puncture incision is used to create a portal through which an endoscope is introduced. The aim of these procedures is to transect 80% of the medial aspect of the plantar fascia. The reason for performing a partial fasciotomy is to avoid injury to the nerve supplying the abductor digiti minimi muscle and to circumvent the potential deleterious effects on the longitudinal arch of the foot that sometimes occurs with a complete transection.

The plantar fascia does not resume its native appearance after a surgical fasciotomy [41]. Approximately one fifth of people who have a surgical release of their fascia demonstrate a persistent gap at the fasciotomy site [Fig. 9]. In a study of 16 asymptomatic people who underwent fasciotomy, the most common postoperative MR imaging appearance 1 year later was a thickened fascia with indistinct margins [42]. On average, it was noted that the thickness of the plantar fascia was 2 to 3 times normal. The most notable observation was complete absence of edema in the fascia or surrounding tissues [42]. Residual areas of intermediate signal intensity on proton density-weighted images were common indicating degenerative changes in the fascia. There was no difference in the appearance of the fascia between an open and endoscopic fasciotomy in this study [42].

Long-term results of plantar fasciotomy indicate that a certain proportion of patients return with persistent or recurrent foot pain. A recent study suggests that these patients can be subdivided

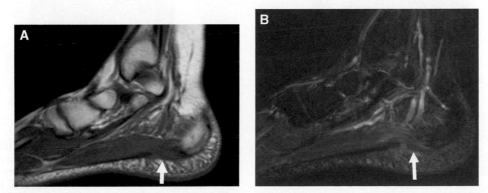

Fig. 9. Plantar fasciotomy. A 50-year-old asymptomatic man who has a history of plantar release. (*A*) Sagittal T1-weighted (TR/TE 500/22) and (*B*) STIR (TR/TE/TI 4000/30/15) images show surgical defect at site of plantar fasciotomy (*arrow*) near the calcaneal origin.

into three broad groups. The most common are those who have persistent or recurrent fasciitis. MR findings vary from plantar fascial thickening and interfascial and perifascial edema to more subtle perifascial edema only. Another group relates to midfoot instability. The plantar fascia normally carries approximately 14% of the total static load. A 10% reduction of the dynamic loading capacity on the foot may occur after plantar fasciotomy. When the height of the longitudinal arch decreases, it accentuates loading on the PTT and the peroneus brevis and longus tendons. When the arch height decreases, two important effects on the foot include an increased incidence of tears of the tendons that support the arch and an increased incidence of midtarsal joint degeneration [43]. The third most common cause of recurrent pain is an acute rupture of the plantar fascia at or near the fasciotomy [42,44].

Tarsal tunnel surgery

Tarsal tunnel syndrome is a compression neuropathy of the posterior tibial nerve as it passes through the fibroosseous tunnel deep to the flexor retinaculum and posterior and inferior to the medial malleolus. Within 1 cm of the medial malleolus, the posterior tibial nerve trifurcates into the medial and lateral plantar nerves and sensory calcaneal branches. Causes of compression neuropathy include lipomas, varicose veins, ganglia, neurilemmomas, scarring, tenosynovitis, and accessory muscles [45,46]. If conservative treatment is not successful, surgical decompression is performed, with division of the retinaculum and mobilization

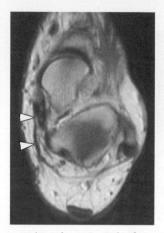

Fig. 11. Flexor retinacular regrowth after tarsal tunnel release. A 45-year-old man who has previous history of tarsal tunnel release presents with recurrent symptoms. Axial T1-weighted (TR/TE 500/22) image shows regrowth of the flexor retinaculum along the medial aspect of the tarsal tunnel (*arrowheads*), causing impingement on the neurovascular bundle.

of the medial and lateral plantar nerves and the fibrous origin of the abductor hallucis [47,48].

Pfeiffer and Cracciollo report that the clinical outcome of tarsal tunnel decompression is more successful when a specific lesion is identified near or within the tarsal tunnel at surgery [49]. Surgical decompression in patients who had prior foot surgery, plantar fasciitis, or systemic inflammatory disease had less favorable results compared with patients who had space-occupying lesions [49].

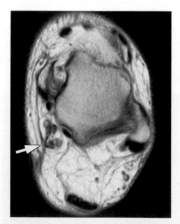

Fig. 10. Tarsal tunnel release. A 35-year-old woman who has a history of tarsal tunnel release. Axial T1-weighted (TR/TE 500/22) image shows scarring in the subcutaneous fat extending to the medial margin of the tarsal tunnel (*arrow*), consistent with prior tarsal tunnel release. The fat otherwise is preserved around the neurovascular bundle.

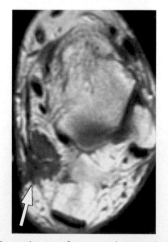

Fig. 12. Scar tissue after tarsal tunnel release. A 48-year-old man who has a history of tarsal tunnel surgery 3 years before MR imaging, now with recurrent symptoms. Axial T1-weighted (TR/TE 500/22) image shows a soft tissue mass (*arrow*) infiltrating the tarsal tunnel with compression of the neurovascular bundle, consistent with scar tissue.

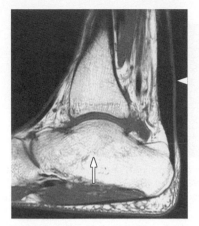

Fig. 13. Flatfoot deformity correction. A 56-year-old man who has surgical correction of flatfoot deformity. Sagittal T1-weighted (TR/TE 500/22) image shows solid subtalar fusion after arthrodesis (*arrow*) and postsurgical thickening (*arrowhead*) of the Achilles tendon, secondary to prior Achilles tendon lengthening.

MR imaging after tarsal tunnel surgery reveals scarring of the subcutaneous fat and fascia along the medial aspect of the foot [Fig. 10] adjacent to the medial malleolus [50]. MR imaging with contrast enhancement can be useful in identifying causes of failed tarsal tunnel release, such as regrowth of the flexor retinaculum [Fig. 11], recurrent ganglia, or fibrous scarring [Fig. 12] impinging on the tarsal tunnel [50,51].

Arthrodesis

The primary indications for ankle arthrodesis are pain, advanced osteoarthritis, and instability. The three surgical principles of ankle arthrodesis are (1) to create congruent cancellous bone surfaces, (2) to stabilize with rigid internal fixation, and (3) to create a plantigrade foot. A plantigrade foot is accomplished by aligning the hindfoot to the lower extremity and the forefoot to the hindfoot.

Ankle arthrodesis includes triple arthrodesis, subtalar arthrodesis, and tibio-talar-calcaneal arthrodesis. A triple arthrodesis consists of the surgical fusion of the talocalcaneal, talonavicular, and calcaneocuboid joints. The primary goal of a triple arthrodesis is to relieve pain from arthritic, deformed, or unstable joints. Other goals include the correction of the deformity and creation of a stable, balanced plantigrade foot. Results of arthrodesis typically are good [52,53].

Subtalar arthrodesis is performed for advanced arthritis of the subtalar joint or for hindfoot deformities, such as pes cavus or clubfoot. Foot deformities often are associated with a tight Achilles tendon. Therefore, subtalar arthrodesis often is accompanied by Achilles tendon surgery, which may involve Achilles tendon lengthening (described previously) or strengthening of the Achilles tendon by tendon transfer procedure [Fig. 13].

Indications for tibio-talar-calcaneus arthrodesis include osteoarthritis, severe ankle trauma, avascular necrosis (AVN) of the talus, failed prior surgery, and rheumatoid arthritis of the ankle and subtalar joints. Because of the complex nature of this surgical technique and the various disease states being treated, complications after arthrodesis are common. These include nonunion, degenerative arthritis, wound dehiscence, sural and superficial peroneal nerve injury, talar AVN, and lateral ankle instability.

In the case of the triple arthrodesis, the talonavicular joint is the most common site of nonunion, with a reported rate of 5% to 10%. A large amount of stress is transferred to the joints immediately distal to the fusion; therefore, midfoot degenerative change can develop with time. Excessive varus or valgus alignment of the hindfoot or forefoot can accelerate the onset of degenerative

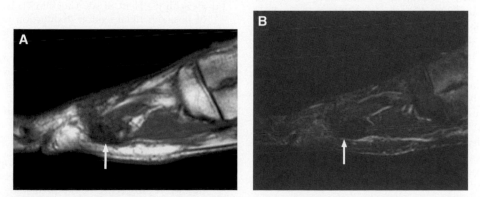

Fig. 14. Metatarsal head AVN after hallux valgus repair. A 62-year-old woman 6 weeks after surgical correction of hallux valgus. (*A*) Sagittal T1-weighted (TR/TE 500/22) and (*B*) STIR (TR/TE/TI 4000/30/15) images demonstrate low marrow signal in the plantar aspect of the head of the first metatarsal, consistent with AVN.

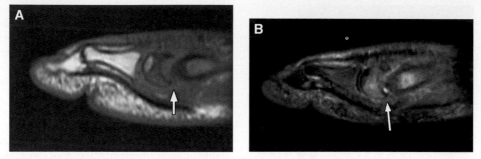

Fig. 15. Nonunion after hallux valgus repair. A 68-year-old woman presents with persistent pain 18 months after hallux valgus surgical repair. (*A*) Sagittal T1-weighted (TR/TE 500/22) and (*B*) STIR (TR/TE/TI 4000/30/15) images show persistent osseous gap and fluid (*arrow*) within the osteotomy site, indicating nonunion with associated pseudoarthrosis.

change [54]. Most complications are identified by radiographs or multidetector CT [54]. MR imaging may be used when greater osseous and soft tissue detail is needed after arthrodesis. It allows accurate assessment of the extent of osseous union, integrity of ankle tendons and ligaments, and detection of postoperative complications, such as infection, nonunion, instability, and secondary degenerative changes.

Hallux valgus repair

There are more than 130 types of surgical procedures described for treatment of hallux valgus. The indication for surgery is persistent pain despite conservative therapy. An osteotomy is a surgical procedure where bone is cut to shorten, lengthen, or change its alignment. Chevron osteotomy is the most common osteotomy of the foot and is an accepted surgical treatment for mild and moderate hallux valgus. In this procedure, a V-shaped osteotomy of the distal metatarsal is created. This allows the first metatarsal head to be shifted laterally, correcting the abnormal shape from longstanding valgus drift. Associated postoperative complications that can be diagnosed at MR imaging include AVN [Fig. 14] or nonunion [Fig. 15]. Distal osteotomies, such as the chevron procedure, are performed in the metaphyseal region. Complications include shortened first metatarsals, second metatarsalgia, restriction of motion, and recurrence of deformity. Diaphyseal osteotomy, such as the Scarf procedure, is associated with decreased blood supply and AVN but can achieve greater correction of deformity. It can lengthen the first metatarsal after rotation or translocation and correct the intermetatarsal angle. Common base or proximal metatarsal osteotomies are the least stable, with the greatest risk for metatarsal elevation and fixation failure.

In conclusion, although routine radiography is the most common imaging modality in assessing the postoperative ankle and foot, MR imaging, with its inherent excellent tissue contrast, has a unique role in evaluating the ligaments and tendons after surgery. It can assess the integrity of these structures and evaluate for postoperative complications. It also has an important role in evaluating the articular cartilage. Awareness of the surgical procedure, the normal postoperative MR imaging appearance, and potential complications is essential for accurate assessment of the postoperative ankle and foot.

Acknowledgments

We gratefully acknowledge Fred Ross for his illustrations.

References

[1] Colville MR. Surgical treatment of the unstable ankle. J Am Acad Orthop Surg 1998;6:368–77.
[2] Platt MA. Tendon repair and healing. Clin Podiatr Med Surg 2005;22:553–60.
[3] Redfern D, Myerson M. The management of concomitant tears of the peroneus longus and brevis tendons. Foot Ankle Int 2004;25:695–707.
[4] Fleischli JG, Fleischli JW, Laughlin TJ. Treatment of posterior tibial tendon dysfunction with tendon procedures from the posterior muscle group. Clin Podiatr Med Surg 1999;16:453–70.
[5] Knobloch K, Kraemer R, Lichtenberg A, et al. Achilles tendon and paratendon microcirculation in midportion and insertional tendinopathy in athletes. Am J Sports Med 2005;34:1–6.
[6] Khan RJ, Fick D, Keogh A, et al. Treatment of acute Achilles tendon ruptures. A meta-analysis of randomized, controlled trials. J Bone Joint Surg [Am] 2005;8-A7:2202–10.
[7] Cetti R, Christensen SE, Ejsted R, et al. Operative versus nonoperative treatment of Achilles tendon rupture. A prospective randomized study and review of the literature. Am J Sports Med 1993;21: 791–9.
[8] Haggmark T, Liedberg H, Eriksson E, et al. Calf

muscle atrophy and muscle function after non-operative vs operative treatment of Achilles tendon ruptures. Orthopedics 1986;9:160–4.

[9] Lindhom A. A new method of operation in subcutaneous rupture of the Achilles tendon. Acta Chir Scand 1959;117:261–70.

[10] Fitzgibbons RE, Hefferon J, Hill J. Percutaneous Achilles tendon repair. Am J Sports Med 1993;21:724–7.

[11] Bradley JP, Tibone JE. Percutaneous and open surgical repairs of Achilles tendon ruptures. A comparative study. Am J Sports Med 1990;18:188–95.

[12] Krackow KA, Thomas SC, Jones LC. Ligament-tendon fixation: analysis of a new stitch and comparison with standard techniques. Orthopedics 1988;11:909–17.

[13] van Schie CH. A review of the biomechanics of the diabetic foot. Int J Low Extrem Wounds 2005;4:160–70.

[14] Lyon R, Liu X, Schwab J, et al. Kinematic and kinetic evaluation of the ankle joint before and after tendo Achilles lengthening in patients with spastic diplegia. J Pediatr Orthop 2005;25:479–83.

[15] Willrich A, Angirasa AK, Sage RA. Percutaneous tendo Achilles lengthening to promote healing of diabetic plantar foot ulceration. J Am Podiatr Med Assoc 2005;95:281–4.

[16] Salsich GB, Mueller MJ, Hastings MK, et al. Effect of Achilles tendon lengthening on ankle muscle performance in people with diabetes mellitus and a neuropathic plantar ulcer. Phys Ther 2005;85:34–43.

[17] Fornage BD. Achilles tendon: US examination. Radiology 1986;159:759–64.

[18] Reinig JW, Dorwart RH, Roden WC. MR imaging of a ruptured Achilles tendon. J Comput Assist Tomogr 1985;9:1131–4.

[19] Marcus DS, Reicher MA, Kellerhouse LE. Achilles tendon injuries: the role of MR imaging. J Comput Assist Tomogr 1989;13:480–6.

[20] Karjalainen PT, Aronen HJ, Pihlajamaki HK, et al. Magnetic resonance imaging during healing of surgically repaired Achilles tendon ruptures. Am J Sports Med 1997;25:164–71.

[21] Moller M, Kalebo P, Tidebrant G, et al. The ultrasonographic appearance of the ruptured Achilles tendon during healing: a longitudinal evaluation of surgical and nonsurgical treatment, with comparisons to MRI appearance. Knee Surg Sports Traumatol Arthrosc 2002;10:49–56.

[22] Hatori M, Matsuda M, Kokubun S. Ossification of Achilles tendon—report of three cases. Arch Orthop Trauma Surg 2002;122:414–7.

[23] Baxter DE. Traumatic injuries to the soft tissues of the foot. In: Mann RA, editor. Surgery of the foot. St. Louis: Mosby; 1986. p. 456–81.

[24] Miller SD. Late reconstruction after failed treatment for ankle fractures. Orthop Clin North Am 1995;26:363–73.

[25] Wilson Jr FC, Skilbred LA. Long-term results in the treatment of displaced bimalleolar fractures. J Bone Joint Surg [Am] 1966;48-A:1065–78.

[26] Fleischli JG, Fleischli JW, Laughlin TJ. Treatment of posterior tibial tendon dysfunction with tendon procedures from the posterior muscle group. Clin Podiatr Med Surg 1999;16:453–70.

[27] Bohay DR, Anderson JG. Stage IV posterior tibial tendon insufficiency: the tilted ankle. Foot Ankle Clin 2003;8:619–36.

[28] Karlsson J, Brandsson S, Kalebo P, et al. Surgical treatment of concomitant chronic ankle instability and longitudinal rupture of the peroneus brevis tendon. Scand J Med Sci Sports 1998;8:42–9.

[29] Porter D, McCarroll J, Knapp E, et al. Peroneal tendon subluxation in athletes: fibular groove deepening and retinacular reconstruction. Foot Ankle Int 2005;26:436–41.

[30] Krause JO, Brodsky JW. Peroneus brevis tendon tears: pathophysiology, surgical reconstruction, and clinical results. Foot Ankle Int 1998;19:271–9.

[31] Borton DC, Lucas P, Jomha NM, et al. Operative reconstruction after transverse rupture of the tendons of both peroneus longus and brevis. Surgical reconstruction by transfer of the flexor digitorum longus tendon. J Bone Joint Surg [Br] 1998;80-B:781–4.

[32] Liu SH, Baker CL. Comparison of lateral ankle ligamentous reconstruction procedures. Am J Sports Med 1994;22:313–7.

[33] St Pierre R, Allman Jr F, Bassett III FH, et al. A review of lateral ankle ligamentous reconstructions. Foot Ankle 1982;3:114–23.

[34] Brostrom L. Sprained ankles. V. Treatment and prognosis in recent ligament ruptures. Acta Chir Scand 1966;132:537–50.

[35] Brostrom L. Sprained ankles. VI. Surgical treatment of "chronic" ligament ruptures. Acta Chir Scand 1966;132:551–65.

[36] Brostrom L, Sundelin P. Sprained ankles. IV. Histologic changes in recent and "chronic" ligament ruptures. Acta Chir Scand 1966;132:248–53.

[37] Hamilton WG, Thompson FM, Snow SW. The modified Brostrom procedure for lateral ankle instability. Foot Ankle 1993;14:1–7.

[38] Bjorkenheim JM, Sandelin J, Santavirta S. Evans' procedure in the treatment of chronic instability of the ankle. Injury 1988;19:70–2.

[39] Becker HP, Ebner S, Ebner D, et al. 12-year outcome after modified Watson-Jones tenodesis for ankle instability. Clin Orthop Relat Res 1999;358:194–204.

[40] Snook GA, Chrisman OD, Wilson TC. Long-term results of the Chrisman-Snook operation for reconstruction of the lateral ligaments of the ankle. J Bone Joint Surg [Am] 1985;67-A:1–7.

[41] Yu JS. Pathologic and postoperative conditions of the plantar fascia: review of MR imaging appearances. Skeletal Radiol 2000;29:491–501.

[42] Woelffer KE, Figura MA, Sandberg NS, et al. Five-year follow-up results of instep plantar fascio-

tomy for chronic heel pain. J Foot Ankle Surg 2000;39:218–23.

[43] O'Malley MJ, Page A, Cook R. Endoscopic plantar fasciotomy for chronic heel pain. Foot Ankle Int 2000;21:505–10.

[44] Yu JS, Spigos D, Tomczak R. Foot pain after a plantar fasciotomy: an MR analysis to determine potential causes. J Comput Assist Tomogr 1999; 23:707–12.

[45] Hirose CB, McGarvey WC. Peripheral nerve entrapments. Foot Ankle Clin 2004;9:255–69.

[46] Labib SA, Gould JS, Rodriguez-del-Rio FD, et al. Heel pain triad (HPT): the combination of plantar fascitis, posterior tibial tendon dysfunction and tarsal tunnel syndrome. Foot Ankle Int 2002; 23:212–20.

[47] Kim DH, Ryu S, Tiel RL, et al. Surgical management and results of 135 tibial nerve lesions at the Louisiana State University Health Sciences Center. Neurosurgery 2003;53:1114–24.

[48] Gondring WH, Shields B, Wenger S. An out-comes analysis of surgical treatment of tarsal tunnel syndrome. Foot Ankle Int 2003;24:545–50.

[49] Pfeiffer WH, Cracchiolo III A. Clinical results after tarsal tunnel decompression. J Bone Joint Surg [Am] 1994;76-A:1222–30.

[50] Recht MP, Donley BG. Magnetic resonance imaging of the foot and ankle. J Am Acad Orthop Surg 2001;9:187–99.

[51] Machiels F, Shahabpour M, De Maeseneer M, et al. Tarsal tunnel syndrome: ultrasonographic and MRI features. JBR-BTR 1999;82:49–50.

[52] Winson IG, Robinson DE, Allen PE. Arthroscopic ankle arthrodesis. J Bone Joint Surg [Br] 2005; 87-B:343–7.

[53] Kopp FJ, Banks MA, Marcus RE. Clinical outcome of tibiotalar arthrodesis utilizing the chevron technique. Foot Ankle Int 2004;25: 225–30.

[54] Raikin SM, Schon LC. Arthrodesis of the fourth and fifth tarsometatarsal joints of the midfoot. Foot Ankle Int 2003;24:584–90.

RADIOLOGIC
CLINICS
OF NORTH AMERICA

Radiol Clin N Am 44 (2006) 407–418

Imaging of the Postoperative Spine

Thomas H. Berquist, MD[a,b,*]

- Indications
- Instrumentation
- Postoperative imaging
  Imaging techniques
  Complications

- Summary
- References

Spinal instrumentation techniques have expanded dramatically during the past several decades, but the search for the perfect operative approach and fixation system continues [1]. Fixation devices are designed for the cervical, thoracic, lumbar, and sacral segments using anterior, posterior, transverse, videoarthroscopic and combined approaches [2,3]. In most cases, bone grafting also is performed, because instrument failure occurs if solid bony fusion is not achieved [2].

Imaging plays an important role in pre- and postoperative imaging of patients undergoing spinal operative or instrumentation procedures [1,4–7]. Postoperative evaluation is important for evaluating position of implants, bone graft fusion, and potential complications. It is essential, therefore, for radiologists to be familiar with the different fixation devices, operative approaches, and potential complications of these procedures. Proper use of the appropriate imaging techniques and knowledge of their advantages and limitations are essential for optimal evaluation of patients who have prior spinal instrumentation.

This discussion focuses on indications, operative approaches, and imaging of spinal instrumentation devices and associated complications.

Indications

Operative intervention may be required for scoliosis and other spinal deformities, degenerative disease, spondylolisthesis, trauma, instability, infection, and neoplasm [8–12]. The goal of spinal instrumentation is to provide stability, reduce deformity by restoring or improving anatomic alignment, and reduce pain. Spinal fusion is the operative method of choice for providing pain relief in patients who have low back pain [10].

Spinal instrumentation is performed most commonly to stabilize spinal fractures, correct scoliosis, and treat multiple causes of lumbar pain related to degenerative disease [2,8–10]. Trauma to the spine is common and may result in instability and compromise of the spinal canal. Operative intervention reduces morbidity and improves mobilization and rehabilitation [2,3]. In most cases of spinal trauma, neurologic deficits do not occur. If left untreated, however, progressive compression and kyphotic deformity occur commonly [1,3,7].

Scoliosis may be congenital, idiopathic, or acquired from any number of neurologic or musculoskeletal disorders [Box 1] [1,8,13]. Infantile

a Mayo Clinic College of Medicine, Rochester, MN, USA
b Department of Radiology, Mayo Clinic, Jacksonville, FL, USA
* Mayo Clinic, 4500 San Pablo Road, Jacksonville, FL 32224.
E-mail address: berquist.thomas@mayo.edu

doi:10.1016/j.rcl.2006.01.002

scoliosis is more common in males. A left thoracic curve occurs most frequently. Curves less than 30° usually resolve. Cardiorespiratory compromise is not unusual with significant curves [14]. Congenital scoliosis commonly is caused by vertebral anomalies [see Box 1]. Other organ anomalies also may be evident. Twenty to twenty-five percent of patients have genitourinary anomalies, especially with lumbar involvement. Cardiac anomalies are more common with congenital thoracic scoliosis [1,15].

Juvenile scoliosis is more common in females. Patients present with a left thoracic curve and progression is more common [15]. Adolescent idiopathic scoliosis often is detected during school physical examinations. A right thoracic curve is most common with a tendency to progress during growth spurts [1,15]. A positive family history is evident in 15% to 20% of patients. Adult scoliosis occurs after skeletal maturity. Adult-onset scoliosis may be related to degenerative disease, trauma, osteoporosis, neoplasms, or other acquired diseases.

Pain and instability in the lumbosacral and cervical regions are common with degenerative disease. These disorders include conditions, such as spondylolisthesis, degenerative spondylolisthesis, spinal stenosis, disc disease, and scoliosis [1, 2,10]. Patients undergoing laminectomy for disc

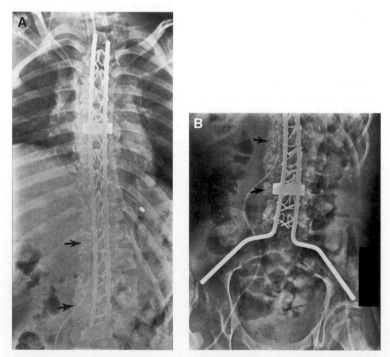

Fig. 1. Spinal fixation with Luque rods. (*A, B*) AP radiographs demonstrate scoliosis corrected with Luque rods from T2 to the sacrum. The rods are attached with sublaminar wires with TSRH cross-links at T6 and L4. There is extensive bone grafting on the right (*arrows*).

disease or spinal stenosis also may require stabilization with instrumentation, depending on the extent of surgery [1,2,10,12,16]. Up to 20% of the population in the United States has low back pain each year; 1% to 2% is disabled temporarily or chronically. Surgery for low back pain continues to increase.

Instrumentation

Procedures for spinal instrumentation may use anterior, posterior, transverse, combined, and, more recently video-assisted arthroscopic surgical approaches. The last can be used for anterior spinal release, bone grafting, and implant insertion [8]. Selection of the approach and instrumentation varies with cause, patient condition, length of fusion required, and surgical preference [1]. Rods, hooks, laminar wires, and cross-links commonly are used for posterior instrumentation in patients who have scoliosis [Fig. 1]. Cancellous screws, cables, and plates may be used for shorter segment anterior instrumentation in patients who have scoliosis.

Posterior instrumentation for degenerative disease in the lumbosacral region is performed commonly with rods, pedicle screws, and posterolateral bone grafting. More recently, interbody fusion cages have gained popularity because of the high failure rates of bone grafting alone and posterior pedicle screw instrumentation alone and the high success rate of anterior interbody fusion cages and bone grafting. Combined anterior and posterior approaches also are used [Fig. 2] [2,3,11]. In the cervical spine, bone block and anterior plate and screw fixation commonly are used [17].

Similar instrumentation may be used in stabilization of the spine after trauma. Posterior instrumentation is used most commonly for flexion injuries. Multiple normal vertebrae above and below the injury typically are included in the instrumentation for greater stability. Anterior instrumentation may be used for hyperextension injuries [18–20]. Table 1 summarizes commonly used instrumentation devices and indications.

Postoperative imaging

Appropriate sequence and selection of imaging techniques are essential in evaluating position of the instrumentation and bone graft material and evaluating potential complications.

Imaging techniques

Baseline radiographs or CR images are essential for evaluating component position and serve as a starting point for evaluation of future studies, should patients develop symptoms suggesting possible complications. Change in position of instrument failure often is appreciated easily on serial radiographs or CR images [Fig. 3]. Anteroposterior (AP), lateral, oblique, and motion studies (flexion, extension, or lateral bending) images usually are adequate. On occasion, however, fluoroscopically positioned images may provide better alignment of the hardware or osseous structures to identify subtle changes more optimally [1,6].

Radionuclide scans may remain positive for a year or more in the region of the operative bed and instrumentation. Combined technetium and labeled white blood cell studies may be useful for

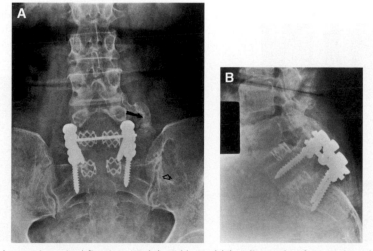

Fig. 2. Anterior and posterior spinal fixation. AP (*A*) and lateral (*B*) radiographs after combined posterior (rod and pedicle screw) and anterior (interbody fusion cages at L4-L5 and L5-S1) fusion for degenerative disease and intractable pain. Note the bone graft donor site (*open arrow*) and posterolateral bone graft (*arrow*) in (*A*). There is no visible bone graft on the right.

Table 1: Spinal instrumentation

Anterior instrumentation

Instrumentation	Indications
Dwyer: cable, staples, cancellous screws	Scoliosis, lordotic deformities
Zielke: rod, staples, cancellous screws	Idiopathic and degenerative scoliosis
TSRH: modification of above	Short segment scoliosis
Kaneda Device: plates, cancellous screws, rods	Trauma
Anterior plates, screws	Trauma, corpectomy
Anterior rods, screws	Trauma, corpectomy
Rezaian system: rib graft and turnbuckle device	Corpectomy
Interbody fusion cages	Degenerative disease, anterior Stability

Posterior instrumentation

Instrumentation	Indications
Harrington system: rods for compression and distraction, hooks	Scoliosis, trauma
Moe and Wisconsin systems: modification of Harrington	Scoliosis, trauma
Luque system: two rods, sublaminar wires	Scoliosis, trauma, congenital deformities
Cotrel-Dubousset: diamond knurled rods, hooks, cross-links	Scoliosis, trauma, spondylolisthesis
TSRH system: rods, screws, hooks, cross-links	Scoliosis, trauma, spondylolisthesis, degenerative disease
ISOLA system: rods, hooks, plates, cross-links, screws	Scoliosis, trauma, tumors, kyphosis, spondylolisthesis, degenerative disease
Rods, plates, pedicle screws	Degenerative disease, spondylolisthesis
Rectangles, sublaminar wires	Trauma, degenerative disease

evaluating infection [1,5]. In recent years, positron emission tomography (PET) has been useful for evaluating infection in the region of metal implants [21].

Ultasound is not used commonly for evaluating potential spinal complications, although detection of superficial abscesses or fluid collections can be accomplished with this technique. CT is useful despite artifact from instrumentation. CT is capable of demonstrating component position, particularly for positioning of pedicle screws [**Fig. 4**] [1]. CT also is useful for evaluating alignment, the spinal canal, and potential complications, such as pseud-arthrosis [1,22,23].

MR imaging has not been used commonly until recently because of artifact from instrumentation devices [**Fig. 5**]. In recent years, titanium implants have been used more commonly, which reduces artifact. Also, multiple imaging parameter changes have been evaluated that reduce artifact and improve image quality. Increasing the frequency encoding gradient strength reduces misregistration

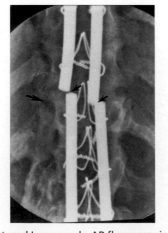

Fig. 3. Fractured Luque rods. AP fluoroscopic spot view demonstrates fractured Luque rods (*arrows*). The bone graft is interrupted at this level (*large arrow*) but intact distally.

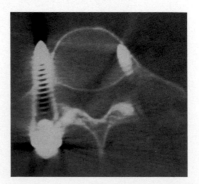

Fig. 4. Pedicular screw malposition. Axial CT image demonstrates breakthrough of the pedicle screw at the base of the pedicle, which extends lateral to the vertebral body.

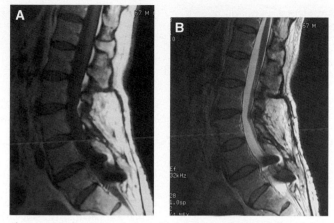

Fig. 5. Laminectomy and posterior instrumentation. Sagittal T1-weighted (*A*) and fast spin-echo T2-weighted (*B*) images after laminectomy and posterior instrumentation with rods and pedicle screws. The postoperative changes and absent spinous processes are obvious. There is artifact from the pedicle screws at the L4-L5 levels. The discs, however, are well demonstrated. Infection could be detected easily if present.

artifact. Orientation of the implant along the magnetic field also reduces the artifact. This is not accomplished easily, however, because of the multidirectional position of the different components. Fast spin-echo sequences increase signal intensity near the implant. Additional parameters, including decreased voxel size, increased bandwidth, and lower field strength, also reduce artifact [1,24].

Fluoroscopically, ultrasonography or CT-guided anesthetic injections are useful for evaluating the source of pain. This may be located at a hook site, facet joint, or disc or in an area of suspected pseudarthrosis. Relief of pain after anesthetic injection confirms the source of pain and allows for proper selection of treatment options. Aspiration of osseous, disc, or soft tissue lesions also is useful when infection is suspected [1]. Table 2 summarizes imaging techniques and indications.

Complications

Imaging plays a vital role in evaluating potential complication of spinal instrumentation procedures. The type of complication varies with the indication for operative intervention, instrumentation, operative approach, patient compliance, and underlying clinical disorders [1,11,12,25]. For purposes of discussion, complications are reviewed based on procedure and types of instrumentation as they relate to scoliosis, degenerative disease, and trauma.

Complications associated with scoliosis instrumentation vary with the type of procedure, patient weight, type of curve, and other clinical factors. Complications may be related to the procedure itself (primary) or to other anatomic regions (secondary) [Table 3]. Certain complications may be increased with operative techniques requiring prolonged immobilization or recumbency (ie, superior mesenteric artery syndrome and thrombophlebitis) [1,26]. When using newer techniques, such as videoarthroscopy, thoracic duct injury and injury to the long thoracic or phrenic nerves may occur [8].

Comparison of pre- and postoperative radiographs or CR images is essential to determine the degree of correction and instrument position. Standing AP, lateral, side-bending, and flexion/extension views are obtained when patients are able to tolerate the examination. These images provide a baseline to measure rotation, curve angles, and kyphosis. The same levels should be selected on each follow-up examination to assure consistency in measurements. Serial radiographs generally are adequate for evaluating loss of correction

Table 2: Imaging of the postoperative spine	
Technique	**Indications**
Radiographs	Instrument failure, infection failed fusion
Fluoroscopic positioned spot views	Instrument position
Radionuclide scans/PET	Infection
Ultrasonography	Fluid collections, abscesses
CT	Instrument position, pseudarthrosis, infection, fragments in spinal canal, vertebral alignment
MR imaging	Infection, failed back, recurrent disc, recurrent tumor
Diagnostic injection/aspiration	Confirm source of pain, Aspiration of pseudobursae or organisms for infection

Table 3: Scoliosis instrumentation complications and imaging techniques

Primary	
Complication	Imaging techniques
Loss of correction	Radiographs
Rod fracture	Radiographs, fluoroscopic spot views
Cable rupture	Radiographs
Hook placement	Fluoroscopic spot views
Screw pull out or fracture	Radiographs
Pseudarthrosis	CT
Neurologic	Radiographs, MR imaging
Infection	Radiographs, CT, MR imaging ultrasound
Local pain	Diagnostic injections
Secondary	
Complication	Imaging techniques
Pneumonia	Radiographs
Pulmonary emboli	CT
Chylothorax	Radiographs, CT
Superior mesentric artery syndrome	MR or CT angiography, conventional angiography
Thrombophlebitis	Ultrasound

Table 4: Lumbar instrumentation complications

Complication	Incidence rates
Pseudarthrosis	5%–32%
Instrument failure	2%–15%
Neurologic	3%–11%
Implant removal	7%–25%
Removal for pain	2%–3%
Iatrogenic flat back syndrome	2%–3%
Infection	1%–2.4%

and instrument failure involving hooks, rods, screws, cables, and other devices [1,13,26]. Fluoroscopic positioning may be necessary to assess the position of the hooks accurately.

Bone grafting is used in conjunction with most instrumentation procedures. It usually takes 6 to 9 months for solid bone graft fusion to be identi-

fied radiographically. CT may be the optimal method for evaluating bone graft using axial and reconstructed coronal or sagittal images. The incidence of pseudarthrosis varies with the type of instrumentation and other patient factors, but it is reported in 15% to 20% of patients. Early recognition of pseudarthrosis or fibrous union is critical to prevent instrument failure and allow early repair [see Fig. 3] [13,25,26]. In difficult cases, diagnostic injections of the suspected area can be accomplished with fluoroscopic or CT guidance [1]. A combination of buffered 1% lidocaine and 0.25% bupivacaine can be used for diagnostic injections. Bupivacaine is longer acting and allows for patient testing to determine the degree of pain relief. Injection of hook sites and facet joints also may be useful for localizing the site of pain and planning appropriate management of the patient [1].

Postoperative infections may be the result of implantation at the time of surgery or occur later in the course of recovery. In the latter situation, the infection may be spread hematogenously or delayed because of seeding at the time of surgery with a latent or subclinical infection [27,28]. The incidence of infection using newer instrumenta-

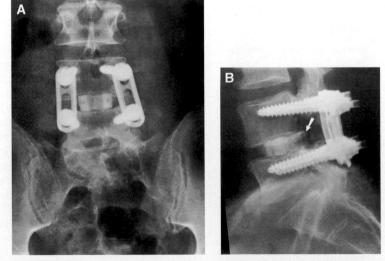

Fig. 6. Displaced interbody bone graft. AP (*A*) and lateral (*B*) radiographs of posterior approach with plate and pedicle screw fusion at L4-L5. The cortical interbody bone graft has displaced posteriorly (*arrow*).

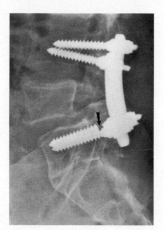

Fig. 7. Pedicular screw fixation. Lateral radiograph after plate and pedicle screw fusion for L4-L5 degenerative subluxation. No attempt is made to reduce the grade 1 subluxation. One of the lower pedicle screws has fractured (*arrow*).

tion (Texas Scottish Rite Hospital [TSRH], Cotrel-Dubousset, or ISOLA) is reported to be 0.6% to 2% [27–29]. Patients present with pain and swelling at the operative site and may develop a draining sinus over time [27]. CT-guided aspiration is useful for isolating the offending organism. Richards [27] reports *Staphylococcus epidermidis* and *Propionibacterium acnes* commonly involved in implant infections. The infection can be treated with antibiotics until healing of the fusion occurs. Removal of the implants should be performed first if the infection is more than 1 year after surgery and the fusion is solid [27–29].

Complications after lumbar or cervical fusion may be related to an underlying degenerative disorder (disc disease, spinal stenosis, spondylolisthesis, or scoliosis), operative approach (anterior, lateral, transverse, posterior, or arthroscopic), and type of instrumentation used [1,10–12,16,30]. For example, there is an increased incidence of fibrosis in the epidural space and spinal canal when using the posterior approach. The incidence of dural tears and retropulsed interbody grafts also is increased [Fig. 6] [11]. Using the laparoscopic approach, iatrogenic disc herniation may be seen as a result of placement of reamers, taps, or fusion cages too far laterally [11].

Arthrodesis of the lumbar spine has evolved dramatically. Initially, posterolateral bone grafting was used. This procedure resulted in a high incidence of pseudarthrosis. Posterior instrumentation or anterior interbody fusion has been added, therefore, to ensure solid fusion [10]. Table 4 summarizes complications of lumbar instrumentation [1,10,11,28,31,32].

A baseline lumbar series should be obtained as soon as patients can tolerate the procedure. Routine studies include AP, lateral, and oblique views. Standing flexion/extension and lateral-bending images may be added to evaluate fusion and changes at adjacent levels. Serial radiographs are useful for evaluating fusion and instrument failure. Detection of infection also is possible, but findings usually are delayed. Additional studies may be needed, when appropriate [see Table 2].

As discussed previously, the addition of posterior instrumentation has improved fusion rates sig-

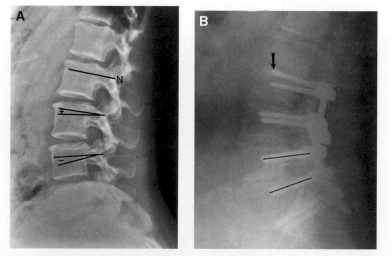

Fig. 8. Normal pedicle screw position. (*A*) Lateral radiograph demonstrates normal screw position. The screw should be centered in the pedicle and aligned neutrally (N). When the screw is angled superiorly, it is considered (+) and inferiorly (–) in position. (*B*) Lateral radiograph with rod and pedicle screw instrumentation from L2-S1. The pedicle screws in L4 and L5 are in neutral position (*lines*). One of the screws in L3 is more inferior, and one in L2 (*arrow*) is eroding into the endplate.

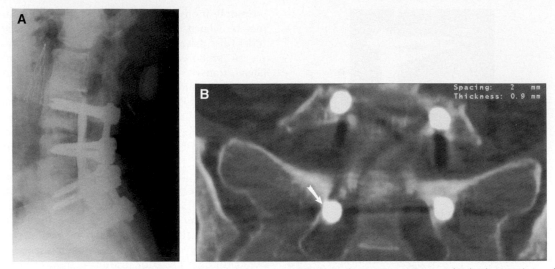

Fig. 9. Malpositioned sacral pedicular screw. (*A*) Lateral myelographic image in a patient who had posterior rod and pedicle screw instrumentation from L3 to the sacrum. (*B*) Coronal reformatted CT image shows the right SI screw entering the neural foramen (*arrow*).

nificantly. Today, pedicle screws with various associated contructs (wire, hook, rod, plate, and cross-links) are used commonly. Several large series review the complications associated with pedicle screws [10,31,33]. Lonstein and colleagues reviewed 915 procedures with insertion of 4790 pedicle screws [32]. There were complications related directly to the screws in 2.4% of screws placed. In this group, 2.8% penetrated the anterior cortex. This usually was not associated with adverse effects. Screws cut out of the cortex during insertion in 1.4% [see Fig. 4]. Pedicle fractures occurred in 0.6% to 2.7% and dural tears occurred in 1% of screw placements. Nerve root irritation was reported in 1% of procedures. This usually was related to medial or inferior placement. Screw fractures occurred in 0.5% of screws placed and 2.2% of procedures [Fig. 7]. The incidence of pseudarthrosis was 65% in 20 patients who had broken screws. The largest concern was late-onset pain resulting from pseudarthrosis or the screws themselves. This requires removal of the instrumentation with or without repair of the pseudarthrosis. Failure of rods, plates, and hooks also may be associated with pseudarthrosis [1].

Screw position and fracture usually can be detected on radiographs. The angle of the screw and portion contained in the vertebral body (50% of the screw length) can be assessed [Fig. 8]. CT, however, is most useful for evaluating position

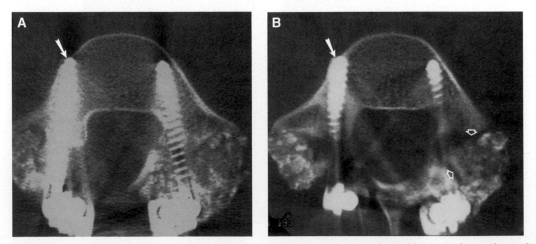

Fig. 10. Axial CT images (*A, B*) after pedicle screw instrumentation and posterolateral bone grafting. The pedicle screw approaches the anterior cortex on the left, and there is slight penetration of the screw on the right. There are no adjacent neurovascular structures. The bone graft is solid and incorporated into the vertebra except at one level (*open arrows*). Coronal reformatting is ideal to evaluate the entire length of the graft.

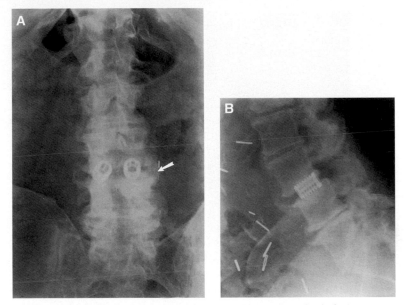

Fig. 11. Interbody fusion cages. AP (*A*) and lateral (*B*) radiographs with interbody fusion cages at L3-L4. The disc space is preserved and there is no lucency at the cage-bone interface. There is bony bridging (*arrow*) laterally on the left.

fully [Fig. 9] and excluding pedicle fracture or cortical penetration [see Fig. 4] [1]. The pedicle screw should be contained within the pedicle and approach the anterior vertebral cortex. CT also is most effective for evaluating the status of bone graft and excluding pseudarthrosis [Fig. 10].

Lumbar interbody fusion is used commonly today. Specific problems with this procedure, therefore, are discussed. Interbody fusion may be accomplished with cortical bone and autogenous grafting or with interbody fusion cages [11,12]. Complications may be related to the device or surgical approach. Retropulsion of the graft or cage is more common with the posterior approach [see Fig. 6] [11].

Serial radiographs may be used to assess fusion. Six criteria are defined by Ray for solid fusion [32]: (1) no motion or less than 3° of intersegment position change on lateral flexion and extension views; (2) lack of a lucent area around the implant; (3) minimal loss of disc height; (4) no fracture of the device, graft, or vertebra; (5) no sclerotic changes in the graft or adjacent vertebra; and (6) visible bone formation in the cage or about the cage [Fig. 11] [6]. Comparing open and laparoscopic procedures, the total complication rate for interbody fusion cages was 14% and 19%, respectively. Many of the complications were minor. Infection was reported in 2% of open and 1% of laparoscopic procedures. Migration of the implant occurred in 1% of open and less than 1% of laparoscopic procedures. Fusion rates of up to 94% were reported 2 years after the procedure [11,34].

Evaluation of interbody fusion may be accomplished more effectively using CT with coronal and sagittal reformatting to evaluate position and fusion. MR imaging is useful for excluding infection and evaluating adjacent discs and vertebral changes. Differentiation of pseudarthrosis from other causes of pain may require discograms, facet injections, and instrument site injections. These techniques are helpful especially before consideration of instrument removal or reoperation [1].

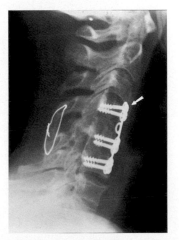

Fig. 12. Combined anterior and posterior instrumentation from C4-C6. The bone blocks are fused. The superior end of the plate extends beyond the level of the vertebral body, with ossification extending over the plate anteriorly (*arrow*).

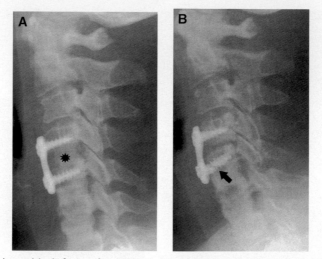

Fig. 13. Prior interbody bone block fusion from C5-C7. (*A*) Immediate postoperative radiograph after fusion of C4-C5 shows the bone graft (*) and Atlantis plate and screws anteriorly. (*B*) Lateral radiograph 1 year later shows partial resorption of the graft with loss of disc height, plate loosening, and fracture of the inferior screw (*arrow*).

Anterior plate and screw instrumentation and interbody bone graft commonly are used to treat cervical degenerative disease [1]. This procedure achieves stability, restores the cervical lordotic curve, and increases fusion rates. Ossification along the anterior longitudinal ligament and adjacent degenerative disc and facet disease, however, occur commonly [1,17]. Ossification of the anterior longitudinal ligament tends to occur more frequently when the end of the anterior plate is too close to the adjacent nonfused disc space. This complication can be avoided by selecting a plate length that results in the end of the plate being equal to or greater than 5 mm from the vertebral margin [Fig. 12] [17].

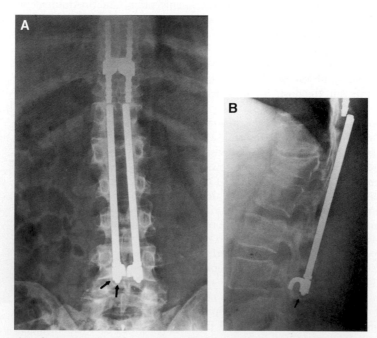

Fig. 14. L1 compression fracture reduced with Harrington distraction rods from T10-L5. AP (*A*) and lateral (*B*) radiographs demonstrate fracture of the upper rods with sclerosis (*arrows*) about the distal rod and hook on the right resulting from motion and loosening. The kyphotic deformity at the fracture site is not corrected.

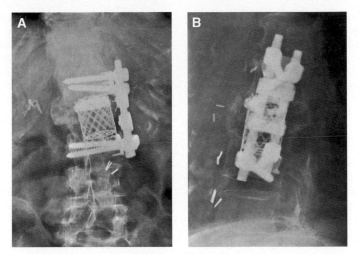

Fig. 15. Anterior fusion, corpectomy, and cage for compression fracture. AP (*A*) and lateral (*B*) radiographs demonstrate anterior fusion using a Kaneda device and titanium Moss cage after corpectomy for a retropulsed L2 compression fracture.

Imaging of cervical fusions is accomplished in the same fashion as lumbar fusion. Again, serial radiographs are the most useful [Fig. 13]. Additional studies can be added as indicated [1].

Complications related to instrumentation after trauma are similar to those discussed previously. The extent of injury, location, and operative treatment all play a role in potential complications. Improper implant selection can result in suboptimal results. For example, Luque rods do not counteract axial loads, which may result in failure to correct vertebral height. Harrington rods used alone may not correct kyphotic deformity created by vertebral compression fractures [Fig. 14] [1,32,35,36].

Location and operative procedure influence the type of complication that may be expected. Anterior procedures in the cervical spine are associated with transient nerve palsies related to retraction during surgery. Vertebral and carotid artery tears, esophageal tears, and bone graft extrusion also are reported [36]. Posterior cervical fusions are associated more often with pseudarthrosis and infection [1,17,36].

Anterior procedures in the thoracolumbar spine have a higher incidence of local complications locally and at the bone graft donor site [Fig. 15]. Vascular injuries also are more common [37]. Neural injury is more common with posterior procedures (cord) and nerve root injury more common with anterior procedures. Pseudarthrosis from instrument failure is more common with long segment fusions and seen more frequently in older patients.

Medical complications also are common. Genitourinary infections occur in 20% and deep venous thrombosis in 25% of patients. These problems are more common in patients who have post-traumatic paralysis or prolonged hospitalization [38]. Gastrointestinal hemorrhage may reach as high as 40% in patients receiving steroid therapy [37].

Summary

In conclusion, it is essential for radiologists to understand the operative and instrumentation options available to surgeons. Knowledge of the expected results, appearance of graft material, and different forms of instrumentation is critical in evaluating position of implants and potential complications associated with operative approaches and spinal fixation devices used.

References

[1] Berquist TH, Currier BL, Broderick DF. The spine. In: Berquist TH, editor. Imaging of orthopedic appliances and prostheses. New York: Raven Press; 1995. p. 109–215.

[2] Slone RM, MacMillan M, Montgomery WJ, et al. Part 2: fixation techniques and hardware for the thoracic and lumbosacral spine. Radiographics 1993;13:521–3.

[3] Krag MH. Spinal fusion. Overview of options and posterior internal fixation devices. In: Frymore JW, editor. The adult spine. New York: Raven Press; 1991. p. 1919–45.

[4] Hayes MA, Tompkins SF, Herndon WA, et al. Clinical and radiological evaluation of lumbosacral motion below fusion levels in idiopathic scoliosis. Spine 1988;13:1161–7.

[5] Mandell GA, Harke HT. Scintigraphy of spinal disorders in adolescents. Skeletal Radiol 1993;22: 393–401.

[6] Whitecloud TS, Skalley T. Roentgen measurement of pedicle screw penetration. Clin Orthop 1989;245:57–68.

[7] Wood KB, Khanna G, Vaccaro AR, et al. Assessment of thoracolumbar fracture classification systems as used by multiple surgeons. J Bone Joint Surg [Am] 2005;87-A:1423–9.

[8] Crawford AH. Anterior spine surgery in thoracic and lumbar spine: endoscopic techniques in children. J Bone Joint Surg [Am] 2004;86-A:2752–63.

[9] Peterson L-E, Nachemson AL. Prediction of progression of the curve in girls who have adolescent idiopathic scoliosis of moderate severity. J Bone Joint Surg [Am] 1995;77-A:823–7.

[10] Hanley EN, David SM. Lumbar arthrodesis for the treatment of back pain. J Bone Joint Surg [Am] 1999;81-A:716–30.

[11] McAfee PC. Interbody fusion cages in reconstructive operations on the spine. J Bone Joint Surg [Am] 1999;81-A:859–78.

[12] Oxland TR, Hoffer Z, Nydegger T, et al. A comparative biomechanical investigation of anterior lumbar interbody cages: central and bilateral approaches. J Bone Joint Surg [Am] 2000; 82-A:383–92.

[13] Yazici M, Asher MA, Hardacker JW. The safety and efficacy of Isola-Galveston instrumentation and arthrodesis in the treatment of neuromuscular spinal deformities. J Bone Joint Surg [Am] 2000;82-A:524–42.

[14] Kostuik JP. Adult scoliosis. In: Rothman RH, Simeone FA, editors. The spine. 3rd ed. Philadelphia: W.B. Saunders; 1992. p. 879–911.

[15] Lenke LG, Bridwell KH, Baldus C, et al. Cotrel-Dubousset instrumentation of adolescent idiopathic scoliosis. J Bone Joint Surg [Am] 1992; 74-A:1056–66.

[16] Stewart G, Sachs BL. Patient outcomes after reoperation of the lumbar spine. J Bone Joint Surg [Am] 1996;78-A:706–12.

[17] Park J-B, Cho Y-S, Riew KD. Development of adjacent-level ossification in patients with an anterior cervical plate. J Bone Joint Surg [Am] 2005;87-A:558–63.

[18] Dunn HK. Anterior stabilization of thoracolumbar fractures. Clin Orthop 1984;189:116–24.

[19] McAfee PC. Spinal instrumentation for thoracolumbar fractures. In: Rothman RH, Simeone FA, editors. The spine. 3rd ed. Philadelphia: W. B. Saunders; 1992. p. 1135–65.

[20] Haas N, Blauth M, Tscherne H. Anterior plating in thoracolumbar spine injuries. Indications, technique and results. Spine 1991;16:S100–11.

[21] Guhlmann A, Breckey-Krause D, Suger G. Chronic osteomyelitis: detection with FDG PET and correlation with histopathologic findings. Radiology 1998;206:749–54.

[22] Golimbu C, Firooznia H, Raffia M. Computed tomography of thoracic and lumbar spine fractures that have been treated with Harrington instrumentation. Radiology 1984;151:731–3.

[23] White RR, Newberg A, Seligson D. Computerized tomographic assessment of traumatized dorsolumbar spine before and after Harrington instrumentation. Clin Orthop 1980;146:150–6.

[24] Peh WCG, Chan JHM. Artifacts in musculoskeletal magnetic resonance imaging: identification and correction. Skeletal Radiol 2001;30:179–91.

[25] Slone RM, MacMillan M, Montgomery WJ. Spinal fixation. Part 3: complications of spinal instrumentation. Radiographics 1993;13: 797–816.

[26] Foley MJ, Calenoff L, Hendryx RW, et al. Thoracic and lumbar spine fusion: post-operative radiographic evaluation. AJR Am J Roentgenol 1983;141:373–80.

[27] Richards S. Delayed infections following posterior spinal instrumentation for treatment of idiopathic scoliosis. J Bone Joint Surg [Am] 1995; 77-A:524–9.

[28] Guedera K, Hooten J, Weatherly W, et al. Cotrel-Dubousset instrumentation. Results in 52 patients. Spine 1993;18:427–31.

[29] Shufflebarger HL, Thomson J, Clark CE. Complications of C.D. in idiopathic scoliosis. Orthop Trans 1992;16:155–6.

[30] Carpenter CT, Dietz JW, Leung KYK, et al. Repair of a pseudarthrosis of the lumbar spine. J Bone Joint Surg [Am] 1996;78-A:712–20.

[31] Yahiro K. Comprehensive literature review. Pedicle screw fixation devices. Spine 1994;19(Suppl 20): S2274–8.

[32] Lonstein JE, Denis F, Perra JH, et al. Complications associated with pedicle screws. J Bone Joint Surg [Am] 1999;81-A:1519–29.

[33] Ray CD. Threaded fusion cages for interbody fusion: an economic comparison with 360 degree fusions. Spine 1997;27:681–5.

[34] Levine AM, Edwards CC. Complications of treatment of acute spinal injury. Orthop Clin North Am 1986;17:183–203.

[35] Luque ER, Cassis N, Romirez-Wiella G. Segmental spinal instrumentation in the treatment of fractures of the thoracolumbar spine. Spine 1982; 7:312–7.

[36] Grob D. Internal fixation of the cervical spine. Spine 1991;16:281–301.

[37] Haas N, Blauth M, Tscherne H. Anterior plating in thoracolumbar spine injuries. Indications, technique and results. Spine 1991;16:S100–11.

[38] Levine AM. Surgical techniques for thoracic and lumbar trauma. In: Rothman RH, Simeone FA, editors. The spine. 3rd ed. Philadelphia: W.B. Saunders; 1992. p. 1104–33.

RADIOLOGIC
CLINICS
OF NORTH AMERICA

Radiol Clin N Am 44 (2006) 419–437

ELSEVIER
SAUNDERS

Imaging of Joint Replacement Procedures

Thomas H. Berquist, MD[a,b],*

- Indications and patient considerations
- Preoperative imaging
- Postoperative imaging
- Complications
- Complications of knee replacement procedures
- Hip replacement complications
- Summary
- References

Hip arthroplasty techniques have progressed dramatically during the past 30 years, providing valuable data for development of joint replacements for other areas [1]. Today, implants are used most commonly for joint replacement of the hip and knee [2,3]. Joint arthroplasty procedures also are performed commonly, however, on the shoulder, elbow, ankle, hand and wrist, and foot [1,4].

The goals of joint replacement arthroplasty are to relieve pain and improve function and quality of life [1,3]. This discussion focuses on indications, pre- and postoperative imaging, and complications associated with joint implants. Clinical and imaging data are presented. Although all joint implants are included, the focus is on joint replacement procedures for the hip and knee.

Indications and patient considerations

Indications for joint replacement procedures vary with anatomic location, patient status, and the underlying disorder. Surgical preference for type and timing of procedures also is a consideration. Most procedures are performed to relieve pain and improve function secondary to arthropathies. Indications for joint replacement include osteoarthritis,

post-traumatic arthropathy, rheumatoid arthritis, other erosive arthropathies, avascular necrosis, and congenital deformities [1–4]. Customized implants may be used for limb salvage procedures in patients who have neoplasms. Contraindications or relative contraindications for joint replacement include vascular insufficiency, active infection, paralysis, severe obesity, and prior arthrodesis [1,5,6].

Patient factors are important for appropriate selection of implants and timing of elective joint replacement procedures. Patient factors include systemic disorders, age, gender, weight, activity level, and patient compliance and expectations.

Age is an important factor. When possible, especially in the hip and knee, joint replacement is reserved for patients 65 years of age or older. Typically, patients are considered candidates if pain interferes with sleep or activity and has not responded to 3 to 6 months of conservative therapy (ie, anti-inflammatory medications or steroid injections). Younger patients (55–65 years of age) are given a longer course of conservative treatment. Patients less than 55 years of age should have more significant symptoms if joint replacement is a consideration. Rand and colleagues report 83% survivorship of knee implants at 10 years for patients

[a] Mayo Clinic College of Medicine, Rochester, MN, USA
[b] Department of Radiology, Mayo Clinic, Jacksonville, FL, USA
* Mayo Clinic, 4500 San Pablo Road, Jacksonville, FL 32224.
E-mail address: berquist.thomas@mayo.edu

radiologic.theclinics.com   doi:10.1016/j.rcl.2006.01.005

Table 1: Anteroposterior view of the pelvis—preoperative image features

Image feature	Indication
Ischial tuberosity line	Leg-length discrepancy Baseline for acetabular angle
Ilioischial line (Kohler's)	Acetabular prostrusio
Femoral neck angle	Component selection
Femoral offset	Component selection
Femoral calcar to canal isthmus ratio	Component selection

less than 55 years of age compared with 94% for patients greater than 70 years old at the time of surgery [5]. Berry and coworkers find Charnley hip implants survived 25 years in 87% of patients implanted at 60 to 70 years of age and in 93% of patients 70 to 79 years of age but only 62% in patients implanted between 40 and 49 years of age [6].

Gender also has an impact on outcomes of joint replacement procedures. Ten-year survival of knee implants is 93% in females and 88% in males [6]. Kurtz and colleagues reviewed revision arthroplasty in the hip and knee over a 13-year period. Revision rates for females were 30% higher than for males [7].

Patient weight is a significant factor. Severe obesity is, at the very least, a relative contraindication for joint replacement. Weight issues are implicated in reduced survival of implants. For example, there is a significant association between acetabular component failure and patient weight exceeding 82 kg (180 lb).

The indication for joint replacement also may have an on impact success rates. Patients who have rheumatoid arthritis treated with Charnley implants have a significantly higher survival rate (92%)

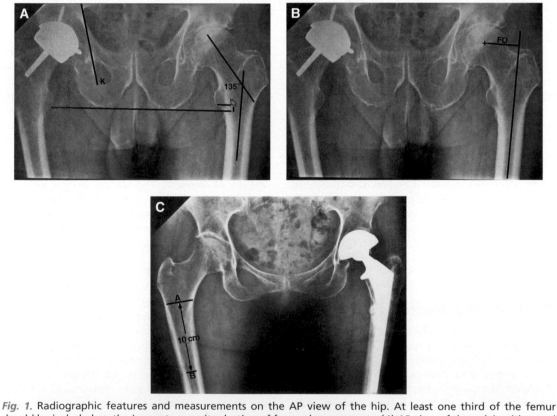

Fig. 1. Radiographic features and measurements on the AP view of the hip. At least one third of the femur should be included on the image to permit selection of femoral components. (*A*) AP view of the pelvis with metal on metal resurfacing implants on the right. The ischial tuberosity line (I) shows slight leg length discrepancy (*open arrow*) where the line intersects the lesser trochanters. Kohler's line (K) shows no protrusion. The femoral neck angle is normal (135°). (*B*) Femoral offset (FO) is measured by the distance from the center of the femoral head (+) to a line centered along the femoral shaft. (*C*) The calcar to canal isthmus ratio is accomplished by measuring the width of the canal at the level of the mid lesser trochanter (A) and 10 cm distally (B). B/A × 100 = 50% normally. If the ratio is 75%, an uncemented porous coated implant may not be useable. Note the total joint replacement on the left.

at 25 years than patients who have congenital hip dysplasia (66%). If considering all indications other than rheumatoid arthritis, the 25-year survival rates are 80% [8].

Preoperative imaging

Clinical and radiographic factors are important for patient and component selection. Clinical scores are established preoperatively to provide a baseline for evaluating postoperative improvement. These scoring systems vary with each articulation. In gen-

eral, however, patients are graded based on scores for pain, function, range of motion, joint stability, and the degree of joint deformity. Pre- and post-operative scores can be compared with assist in documenting the degree of improvement [1].

Multiple imaging modalities may be required to evaluate osseous, articular, and soft tissue structures fully. Routine radiographs or computed radiography (CR) images are necessary for measuring angles and templating for implant size and selection. Magnification markers are placed by the patient during generation of the images to assure accurate mea-

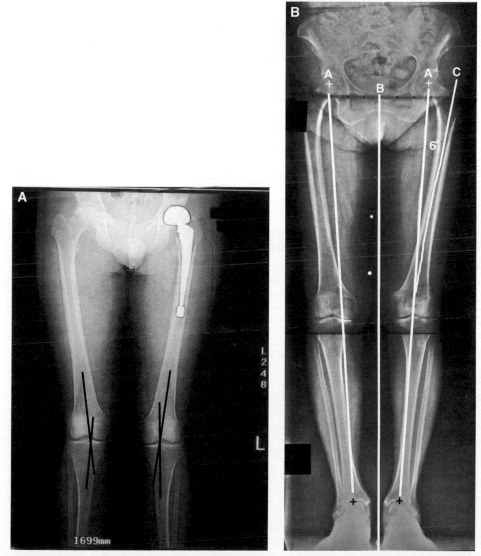

Fig. 2. Preoperative leg radiographs. (*A*) CT scanogram demonstrates femorotibial angles measurements using lines through the centers of the femurs and tibiae. Both are 5°–7° valgus. (*B*) Full-length standing AP radiograph. The mechanical axis (A) is a line from the center of the femoral head (+) through the knee to the center of the ankle (+). This normally passes through the center of the knee (as on the left). The right mechanical axis passes medial to the joint, due to osteoarthritis and medial joint space narrowing. The vertical axis (B) is a straight line from the center of the symphysis. The femoral axis (C) is approximately 6° valgus to the mechanical axis (A).

surements on the images. Radiographs also are useful to evaluate bone stock, joint changes, and ensethopathy (diffuse idiopathic sclerosing hyperostosis [DISH]) [1,9,10]. Anteroposterior (AP) and lateral views usually are adequate for preoperative evaluation. In some cases, additional views or stress views may be required. Standing full length (above the hip to below the ankle) images are obtained before knee replacement [1].

CT is useful for evaluating bone changes, such as subchondral cystic changes or prior operative defects [1,11,12]. On occasion, MR imaging or MR arthrography may be useful to evaluate articular cartilage, avascular necrosis, or soft tissue support [1].

Although similar features are evaluated for all joints, this article focuses on the hip and knee. Table 1 lists radiographic features and indications evaluated on AP preoperative images of the hip. Oblique and lateral views also are evaluated. Imaging of the knee includes AP, lateral, standing flexion (joint space measured more accurately), patellar (Merchant view), and full-length standing AP images. The full-length standing view permits evaluation of hip, knee, and ankle articulations; leg length discrepancy; osseous deformities; and several important measurements. Figs. 1 and 2 summarize features assessed in the hip and knee, respectively.

Postoperative imaging

Radiographs or CR images are essential for postoperative evaluation of component position and serve as a baseline for evaluating complications. Serial images are the most useful tool in follow-up should complications occur. In most cases, radiographs are obtained as soon as patients can tolerate the examination and then at 3, 6, and 12 months. Thereafter, imaging is obtained as indicated.

Baseline images of the knee and shoulder should be positioned fluoroscopically to ensure optimal alignment of the component-cement or component-bone interfaces. Proper visualization of the interfaces and component position is difficult for the tibial, patellar, and glenoid components using conventional positioning techniques [1]. Most other joint replacements can be evaluated using conventional imaging views [Table 2].

It is important to understand the proper position of components regardless of the joint replacement procedure. It is beyond the scope of this discussion to include all joints. Therefore, this article focuses on component position in the hip and knee. AP and lateral views of the hip [see Table 2] generally are adequate to evaluate the acetabular and femoral component positions [Fig. 3]. The AP view should include the entire femoral stem, cement that ex-

Table 2: Postoperative imaging of joint replacements

Articulation	Radiographic views
Hip	AP and lateral
Knee	Standing AP, full length standing AP, Merchant, lateral[a]
Ankle	AP, lateral, mortise flexion/extension lateral
Foot	AP, lateral, oblique
Shoulder	AP, scapular Y, axillary[a]
Elbow	AP, lateral
Wrist	AP, lateral, both obliques
Hand	AP, lateral, oblique

[a] Fluoroscopic positioning preferred for first baseline study.

tends beyond the component, cement restrictor, and several centimeters of bone below the construct. Several of the preoperative measurements [see Table 1] are repeated, including Kohler's line and the ischial tuberosity line. The acetabular component should be angled approximately 45° (range, 35°–55°) [see Fig. 3]. The femoral component should be positioned symmetrically in the acetabular cup. A line from the center of the femoral head or acetabular articular margin to Kohler's line can be used as a baseline to evaluate medial migration of the acetabular component should it occur later. On the lateral view, the acetabular component should be anteverted approximately 15°. Some surgeons may place the acetabular component at a slightly greater angle. The femoral component should be aligned with the femoral shaft or slightly valgus [see Fig. 3]. Varus position is not optimal [1]. Evaluation of the cement and osseous structure about the components are discussed later in the complication section.

In the knee, the mechanical axis; femorotibial angle; and tibial, patellar, and femoral component positions are evaluated using fluoroscopically positioned images or full-length standing AP, standing AP, and lateral and patellar views [Fig. 4]. The tibial tray should cover at least 85% of the surface of the tibia. It can overhand laterally to some degree. Medial overhand of the tibial tray, however, may lead to pes anserine bursitis [see Fig. 4]. The tibial tray should be 90° to the tibial shaft on the AP and lateral views [Fig. 5]. On the AP view, if the angle is greater than 90°, it is considered valgus, and if it is less than 90°, varus [see Fig. 5] [13]. The femoral component should be 97° to 98° related to the femoral shaft on the AP view. The femorotibial angle should be 5° to 10° valgus. On the lateral view, the femoral component should be perpendicular to the femoral shaft [see Fig. 5]. The patellar component should be centered between the con-

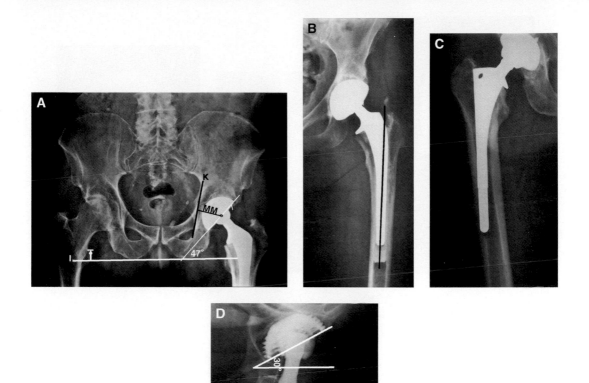

Fig. 3. Postoperative evaluation of total hip replacement. (*A*) AP view of the pelvis. The femoral component is not visualized completely. The ischial tuberosity line (I) shows slight leg length discrepancy (*arrow*). The acetabular component is angled 47° (normal range, 35°–55°). A line (MM) from the center of the femoral head to Kohler's line (K) can be used to follow medial migration. (*B*) The femoral stem (*line*) is just off neutral position in relation to the femoral shaft. Neutral to slight valgus is preferred. (*C*) Uncemented DePuy right hip replacement with the femoral stem is shifted into varus and erodes through the cortex as a result of loosening. (*D*) Lateral view of the hip with a threaded acetabular component and ceramic femoral head. The acetabulum is angled 30° (usually neutral to 15°, but some surgeons prefer more anteversion).

dyles of the femoral component without significant tilt [see Fig. 5] [14,15]. In some cases, an oblique posterior condylar view is added to evaluate the posterior condyle better, which can be difficult on conventional lateral views [16].

Complications

Complications related to joint replacement procedures may be related to patient status, surgical expe-

rience, operative approach (anterior, posterior, and so forth), type of implant selected (constrained, unconstrained, cemented, uncemented, hemiarthroplasty, unipolar or bipolar in the hip, single compartment in the knee, and so forth), and length of hospital stay [1,2,17–19]. Early and delayed complications occur. If considering the first 90 days after surgery, the potential complications may be medical or directly related to the implants. Reviewing primary and revision arthroplasties in the knee,

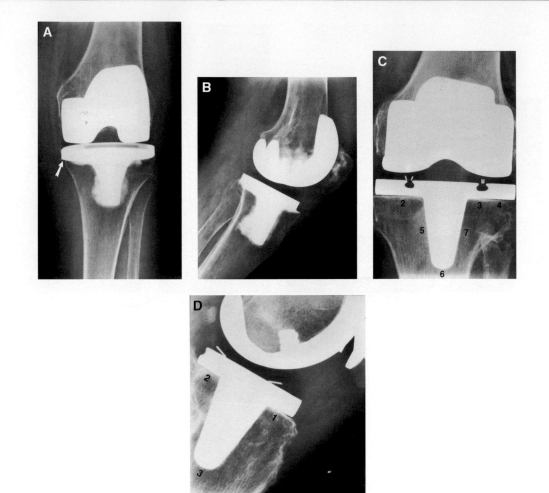

Fig. 4. Postoperative images of the knee. AP (*A*)) and lateral (*B*) radiographs are not aligned well and suboptimally display the prosthetic-cement/bone interfaces. Fluoroscopically positioned AP (*C*) and lateral (*D*) images demonstrate zones for evaluating lucent lines and proper alignment of the prosthesis bone interfaces.

Mahomed and colleagues report mortality rates of 0.7%, readmission rates of 0.9%, pulmonary embolus 0.8%, wound infection 0.4%, pneumonia 1.4%, and myocardial infarction in 0.8% of all patients who underwent unilateral knee replacements in the year 2000 [2]. Some short-term complications were higher after revision arthroplasty, including readmission rates of 4.7% of patients and wound infections in 1.8%. The 30-day mortality rates in 30,714 consecutive hip replacements were 0.29% [18].

Complications also are higher when surgeons perform lower volumes of joint replacement procedures. Hammond and coworkers report complication rates for shoulder arthroplasty of 14.5% with low surgical volumes (1–5/year) compared with 9.3% for high-volume surgeons (greater than 30 cases/year) [17]. Hospital stays also were longer when performed by the low-volume group.

Dislocation of hip components is uncommon (3%). There is a tendency, however, for this complication to occur in the early postoperative period when patients begin weight bearing [Fig. 6] [1]. Fractures may occur during the procedure with reaming or component placement. Fractures also may occur with initial weight bearing or as a late complication [1,20,21].

Delayed complications vary with operative approach, implant selection, anatomic location, and patient factors, including compliance [1,5,8,22]. There are significant data regarding hip and knee replacement. This article focuses, therefore, on complications of these implants. The initial discussion and Tables 3 to 8, however, summarize imaging of complications for the ankle, foot, and upper extremity.

There are several series reviewing ankle joint replacement [4,23–25]. These reviews include Mayo,

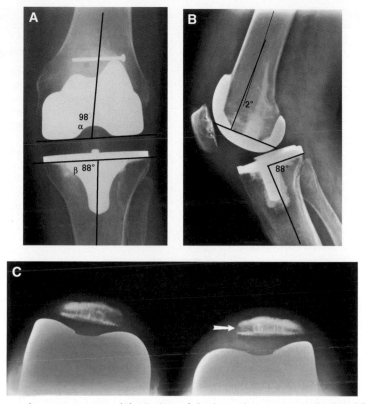

Fig. 5. Postoperative angle measurements. (*A*) AP view of the knee demonstrates tibial and femoral component alignment. The femoral component should be angled (α) 97°–98° to the femoral shaft. The tibial tray should be 90° to the tibial shaft. An angle (β) less than 90° is considered in varus position and an angle greater than 90° valgus. In this case, the angle is 88°. (*B*) Lateral view of the knee. The tibial tray again should be at 90° and in this case it is minimally off at 88°. The femoral component should be perpendicular to the femoral shaft. In this case, it is off by 2°. (*C*) Bilateral all polyethylene patellar components, with symmetric positioning on the left and slight lateral subluxation on the right.

Scandinavian Total Ankle Replacement (STAR), and Agility implants [see Table 3]. Best results with the Mayo ankle arthroplasty (204 patients) occurred in patients who had rheumatoid arthritis and patients who had posttraumatic arthro-

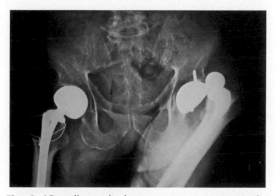

Fig. 6. AP radiograph demonstrates posterosuperior dislocation of the left hip that occurred during initial weight bearing.

sis who were over 60 years of age. Survival rates were 79% at 5 years, 65% at 10 years, and 61% at 15 years [23].

In a series of 51 patients who had the STAR, 12 patients (23.5%) required revision for loosening or other complications. Ninety-five per cent of the patients followed for 4 years did not require reoperation and were satisfied with results [4].

Reports on the Agility implants included 438 patients. Most procedures (90%) were performed for post-traumatic or osteoarthritis. Failure is reported in 11% of patients reported by Spirt and coworkers, defined as removal or replacement of components, ankle arthrodesis, or below-the-knee amputations [25]. In all series, revision rates range from 11% to 28% [**Fig. 7**] [4,23–25].

Metatarsophalangeal joint replacements are performed infrequently compared with the more proximal joints in the lower or upper extremity. Indications do not differ significantly from joint replacement procedures in other articulations. Procedures are performed most commonly on the

Table 3: Ankle joint replacement complications

Complication (incidence %)	Image techniques
Loosening (14%–24%)	Serial radiographs, CT
Deep infection (1.5%–3%)	Serial radiographs, radionuclide scans, joint aspiration
Wound healing (3.5%–4%)	Usually clinical diagnosis
Subtalar arthrosis (19%)	Serial radiographs, CT
Polyethylene fracture (4%)	Serial radiographs
Tibial component fracture (2%)	Serial radiographs
Impingement (13%–15%)	Serial radiographs, CT

Table 5: Shoulder joint replacement complications

Complications (incidence %)	Imaging techniques
Glenoid component loosening (3%–15%)	Fluoroscopically positioned images
Humeral component loosening (1%–7%)	Serial AP, lateral radiographs
Subluxation/ dislocation (6%)	Serial AP, axillary, and scapular Y views
Superior migration (22%)	Serial AP, scapular Y views
Nerve injury (1%–2%)	MR imaging
Humeral fracture (1.6%)	Serial AP, lateral radiographs

great toe. The lesser metatarsophalangeal joints, however, also are treated with joint replacement. Implants used include double-stemmed siliastic, single-stemmed silastic, and metal and polyethelene components [26,27].

Complications include heterotopic ossification around the implants, loss of motion, recurrent deformity, stress fractures, and deep infection [see Table 4] [1,27]. Serial radiographs are most useful for evaluating complications. CT is helpful for evaluating the bone interfaces about the components. MR imaging also is useful, especially for silastic implants, as there is no artifact. Therefore, if infection is suspected, MR imaging with contrast-enhanced, fat suppressed, T1-weighted images included in the sequences may be optimal for diagnosis.

Shoulder replacement procedures include hemiarthroplasty, total joint replacement, and newer procedures for patients who have irreparable rotator cuff disease. In the last setting, the Delta reverse ball and socket system is used [1,28,29]. Complications with total shoulder replacement [see Table 5], like the hip and knee, include loosening and infection [Fig. 8]. Glenoid component loosening [see Fig. 8] occurs most frequently [1]. Experience with the Delta system is limited. Early studies, however, demonstrate that up to 50% of patients have complications, even though they may be minor in nature. One third of patients reported by Werner and colleagues required re-operation [29].

Elbow replacement complications depend on the type of implant in addition to the other common factors. Components may be semiconstrained (Conrad-Morrey) and linked or unlinked. Older constrained designs are not used commonly today [1,30–32]. In a large series reported by Morrey, infections occurred in 8.1%, loosening in 6.4%,

Table 4: Metatarsophalangeal joint replacement complications

Complication	Imaging techniques
Ossification around implant	Serial radiographs, CT
Implant fracture	Serial radiographs, CT
Loss of motion	Stress views
Recurrent deformity	Serial radiographs
Deep infection	Serial radiographs, radionuclide scans, MR imaging for silastic implants
Stress fractures	Serial radiographs, radionuclide scans
Silicone synovitis	MR imaging

Table 6: Elbow joint replacement complications

Complication (incidence %)	Imaging techniques
Loosening (6.4%–17%)	Serial radiographs
Infection (8.1%)	Serial radiographs, radionuclide scans, aspiration
Instability (7%–19%)	Fluoroscopically positioned stress views
Dislocation (2.2%–4.3%)	Serial radiographs
Fractured components (0.6%)	Serial radiographs
Ulnar nerve lesions	MR imaging (metal artifact may be significant)

Table 7: Wrist and hand joint replacement complications

Complication (incidence %)	Imaging techniques
Loosening (4%–30%)	Serial radiographs, CT
Deep infection (1%–2%)	Serial radiographs, radionuclide scans, MR imaging for silastic implants
Soft tissue imbalance (20%–35%)	Serial radiographs, motion studies
Dislocation (4%–10%)	Serial radiographs
Tendon rupture (6%)	Ultrasound, MR imaging
Nerve compression (2%–3%)	MR imaging, ultrasound
Silicone synovitis	Contrast-enhanced MR imaging

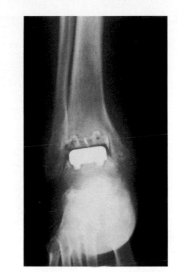

Fig. 7. AP radiograph of the ankle after joint replacement demonstrates lucent lines about the tibial and talar implants resulting from loosening.

ulnar nerve lesions in 10.4%, and instability in 7% to 19% [see Table 6] [31]. Instability and dislocation are a more significant problem with unlinked designs [30].

Wrist arthroplasty is performed most often for rheumatoid arthritis. Individual carpal bone replacement may be performed for traumatic arthrosis, avascular necrosis, and other arthropathies. Silastic implants are used most commonly in this setting. Replacement of the metacarpophalangeal and interphalangeal joints usually is accomplished with single- or double-stem implants. Complications are summarized in Table 7 [1,33]. Once again, MR imaging may be used more commonly to evaluate complications, as there is no artifact with silastic implants [1].

Although complications in the joints that are replaced less frequently are summarized, the major focus is discussing complications after hip and knee replacement procedures.

Complications of knee replacement procedures

Complications after knee replacement are related to component selection, patient selection, and surgical technique [5]. Joint replacement may be limited to a single compartment (typically medial) or may include medial and lateral without patellar resurfacing, and all three compartments may be replaced [5,22,34]. Condylar (posterior cruciate saving) and posterior stabilized (posterior cruciate sacrificed) implants also have an impact on potential complications [1,5,35]. Ten-year survivals for nonmodular tibial trays are 92%; modular tibial trays, 90%; and all polyethylene tibial components, 97%. All polyethylene patellar components survived 10 years in 93% compared with 76% for metal-backed patellar components. Also, 10-year survivals for cemented components were 92% compared with 61% for uncemented components [5].

Although infection and loosening long have been considered the most significant complications, extensor mechanism problems are the number one problem after knee replacement [see Table 8] [1,34–39]. Complications include patellar fracture, subluxation, dislocation, instability, quadriceps rupture, patellar tendon rupture, loosening, syno-

Table 8: Knee replacement complications

Complication (incidence %)	Imaging techniques
Extensor mechanism (4%–41%)	Serial radiographs, flexion/extension axial images
Loosening (2%–5%)	Serial radiographs, radionuclide scans, CT
Deep infection (1%–2.2%)	Serial radiographs, radionuclide scans, joint aspiration
Instability (13%)	Stress views
Fractures (1.2%–3%)	Serial radiographs
Polyethylene wear	Serial radiographs, CT
Osteolysis (7%)	Serial radiographs, CT

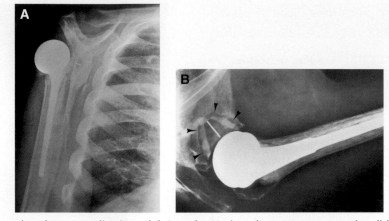

Fig. 8. Shoulder arthroplasty complications. (*A*) Scapular Y view demonstrates posterior dislocation. (*B*) Fluoroscopically positioned image shows wide zones of lucency about the polyethylene glenoid component (*arrowheads*) resulting from loosening.

vial entrapment, and component wear [Fig. 9] [1,34–39]. Synovial entrapment occurs with hypertrophy of synovial tissue proximal to the patella. This complication occurs most commonly with posterior stabilized implants (12% of patellar complications) [35]. Quadriceps rupture is rare (0.1% in 23,800 cases), but treatment results are poor, especially with complete tears [37]. Extensor mechanism complications account for up to 50% of knee revisions [34]. Most complications occur 1 to 2 years after surgery. As discussed previously, success rates are higher with all polyethylene components than with metal-backed components [5].

Serial radiographs (AP, lateral, and Merchant view) usually are adequate for detection of extensor mechanism complications [see Fig. 9; Fig. 10]. Flexion and extension lateral view and axial views in varying degrees of flexion may be useful in subtle cases. MR imaging and CT can be used for soft tissue injury and bone changes, respectively. These techniques are most useful in the presence of polyethylene implants as there is no local artifact [1].

Patients who have loosening generally present with pain. The incidence of loosening was high with early-hinged designs (20%–30%), but this complication has been decreased to a lower percent

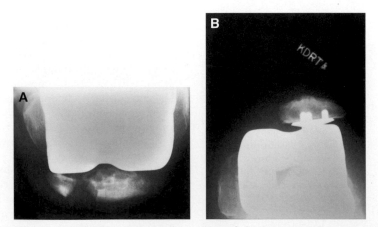

Fig. 9. Extensor mechanism complications. (*A*) Merchant view of the knee demonstrates patellar fracture in a patient who has an all polyethylene component. (*B*) Merchant view demonstrates a subluxed metal-backed patellar component.

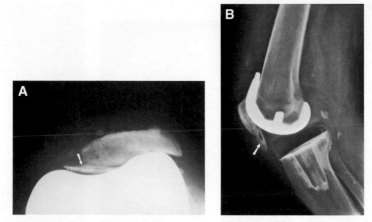

Fig. 10. Loose patellar component. Merchant (*A*) and lateral (*B*) views of the knee demonstrate loosening and displacement (*arrow*) of a metal-backed patellar component.

with newer unconstrained or semiconstrained implants [see Table 8] [1,5,40]. Tibial component failures occur more frequently than femoral component complications [Fig. 11]. When a tibial tray is in greater than 5° of varus, the likelihood of loosening is increased [see Fig. 11] [40]. Evaluation of the component-bone or cement-bone interface can be accomplished with serial radiographs. Subtle lucent zones (less than 2 mm) are common about tibial (65%) and femoral components (20%). Clinical loosening, however, is unlikely unless there is progression or lucent zones greater than 2 mm are demonstrated [Fig. 12] [1,11]. Additional imaging modalities, such as radionuclide scans, CT, and, more recently, MR imaging, can be helpful in selected cases [1].

Superficial or wound infections and deep infections both occur. The former usually are in the immediate postoperative period. The incidence of deep infection is reported at 1.2% to 2% in multiple large series [1,42]. Infections are classified on the basis of clinical presentation: Type I—positive

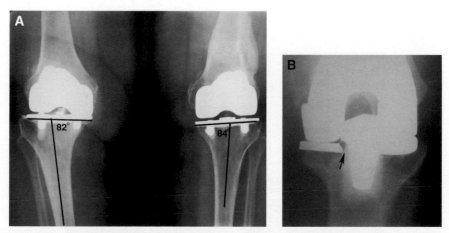

Fig. 11. Tibial component complications. (*A*) Standing AP radiograph demonstrates bilateral Howmedica porus coated anatomic (PCA) implants. The left tibial tray covers the bony surface and is angled at 84° to the tibial shaft (normal, 90° ± 5°). The tibial tray on the right overhangs medially and is angled at 82°, which increases the likelihood of polyethylene wear (note obliteration of the medial spacer) and loosening. Note the old left distal femoral fracture. (*B*) Notch view demonstrates a fracture of the tibial tray (*arrow*).

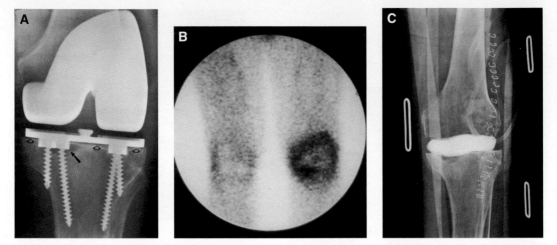

Fig. 12. Infected knee replacement. (*A*) AP radiograph demonstrates lucent lines (*open arrows*) with a larger lucent zone near the medial screws (*arrow*). Is this loosening and infection? (*B*) Indium In 111–labeled white blood cell scan demonstrates intense uptake in the knee. (*C*) Implant was removed and an antibiotic impregnated spacer was inserted.

intraoperative culture; Type II—early postoperative, either superficial or deep; Type III—acute hematogenous; and Type IV—late chronic infections [43]. Surgical intervention is required for débridement or component removal, unless the infection is superficial or there is a positive intraoperative culture. Antibiotic therapy may be successful in these settings [43]. Infections are most commonly the result of *Staphylococcus aureus, Staphylococcus epidermidis, Pseudomonas aeruginosa,* and *Eshcericia coli* [1,43].

Routine laboratory studies and joint aspiration may isolate the offending organism. Joint aspira-

tions are successful in 67% to 90% of cases [1]. Serial radiographs may demonstrate loosening or periosteal changes. Radionuclide scans can be useful in chronic infections (more than 10–12 months after surgery) especially when combined technetium and labeled white blood cell scans or PET images are obtained [see Fig. 12] [1,42].

Polyethylene wear results in joint asymmetry on an AP radiograph [see Fig. 11]. There is frequently associated osteolysis [Fig. 13]. These complications may be evident in up to 7% of cases. Polyethylene wear is more frequent when femoral and tibial

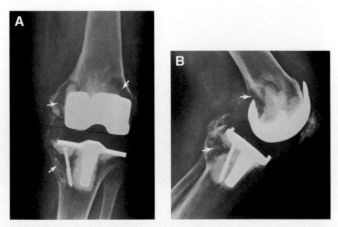

Fig. 13. Particle disease resulting from polyethylene spacer wear. AP (*A*) and lateral (*B*) radiographs of the knee with slight spacer asymmetry (*line*) and extensive osteolysis (*arrows*).

components are not placed perpendicular to the mechanical axis. Patients who are obese or who are overly active also tend to have an increased incidence of polyethylene wear [40].

Fractures of the patella [see Fig. 9], tibia, or femur also may occur. The incidence is reported to range from 1.2% to 3%. Fractures occur more frequently in patients who are osteopenic, patients who have hinged or more constrained components, and with technical problems during surgery. Stress or insufficiency fractures also may occur with ambulation [1]. Fractures may be detectable on radiographs. Subtle fractures can be evident with radionuclide studies or if not immediately adjacent to the metal with MR imaging [1].

Instability occurs in up to 13% of patients depending on preoperative soft tissue status, type of components selected, and patient weight. This diagnosis can be made clinically in most cases. Stress views can supplement the clinical examination when indicated [1].

Other complications include synovitis, which may be the result of metal particles in addition to other causes of inflammation. Pes anserine bursitis can occur with unicompartmental procedures or overhang of the tibial tray medially [see Figs. 4 and 11]. Although uncommon, peroneal nerve palsy also is reported [1,40].

Hip replacement complications

Multiple procedures can be selected for hip replacement including unipolar, bipolar, and total joint arthroplasty. The last may include metal on polyethylene, metal on metal [see Fig. 1], and ceramic components [see Fig. 3] [1,6,8,9,44,45]. Components may be cemented in the acetabulum and femur, femur alone, or both components used without cement. Surgical approaches, implants selected, and patient factors all play a role in the types of complications that may occur [1,9,44]. As in the knee, loosening and infection are significant complications. Multiple complications, however, can occur [Table 9].

Component loosening has been reduced significantly with new components, improved cement techiques, and modified surgical approaches [1,6, 8,10]. Loosening rates have decreased from as high as 57% to 2% to 8% for acetabular components and 5% to 18% for femoral components [Fig. 14] [1,6,8]. Radiographic criteria for cemented acetabular component loosening include lucent zones at the bone cement interface of greater than 2 mm, especially in zone II, progressive widening of the lucent zone, component migration, and cement fracture. With newer uncemented components,

Table 9: Hip joint replacement complications

Complication (incidence %)	Imaging techniques
Loosening Acetabular component (2%–8%) Femoral component (6%–18%)	Serial radiographs, CT, subtraction arthrograms, radionuclide scans
Deep infection (2%–3%)	Serial radiographs, radionuclide scans, joint aspiration, subtraction arthrograms
Dislocation (3%–4.8%)	Serial radiographs
Osteolysis (5%–7%)	Serial radiographs, CT
Pseudobursae (43%)	Subtraction arthrograms
Greater trochanteric avulsion/ nonunion	Serial radiographs
Fractures (1.2%)	Serial radiographs
Heterotopic ossification	Serial radiographs

there are five criteria, including lucent lines developing 2 years after surgery, progression of lucent zones 2 years after surgery, radiolucent lines in all three zones, radiolucent lines greater than 2 mm in any zone, and component migration [see Fig. 14] [10]. The incidence of acetabular component loosening increases over the life of the components from 1.3% at 10 years to 10% at 25 years [6].

Loosening of cemented femoral components also increases over time. Berry and colleagues report loosening in 2.6% at 10 years and 10% at 25 years with Charnley implants [6]. Loosening rates for the femoral component in multiple series with differing components have ranged from 3% to 18% [1,6]. Changes about the femoral component are described in seven zones [Fig. 15]. Radiographic features for loosening of cemented femoral components include varus migration, cement fracture [see Fig. 14], stem fracture, lucent zones at the bone cement interface of greater than 2 mm, and progression of lucent zones. Serial radiographs are useful and accurately predict component loosening of the acetabular component in 69% and femoral component in 84% [1].

Loosening of uncemented femoral components is more difficult to evaluate as lucent zones are seen commonly [see Fig. 15] and the component does not fill the medullary canal entirely. Additional studies, including subtraction arthrography [see

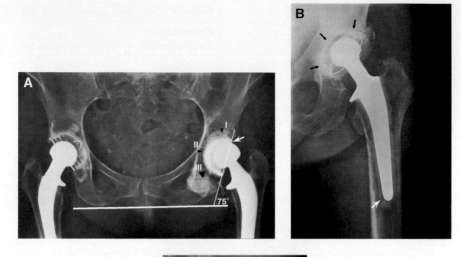

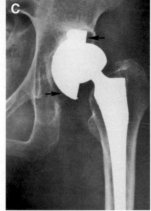

Fig. 14. Acetabular component loosening. There are three zones about the acetabular component (I – III) dividing the region into three areas lateral to medial. (*A*) AP radiograph of the pelvis demonstrates a steep left acetabular angle of 75° (normal = 45°). There are lucent lines at the bone cement interface (*small arrowheads*) in zones I and II, with migration (*large arrowhead*) in zone III. The femoral head is positioned asymmetrically (*arrow*), a result of polyethylene wear. (*B*) AP view of the left hip demonstrates a wide irregular lucent line in all three zones (*arrows*), a result of loosening of the acetabular component. There also is a cement fracture (*white arrow*) at the tip of the femoral component due to loosening. (*C*) Porous coated acetabular component with migration (*arrows* show direction of motion) of the component.

Fig. 15], radionuclide studies, and diagnostic anesthetic injections, are helpful to evaluate cemented and uncemented implants more effectively [1].

Early results after hip arthroplasty demonstrate infection in 8% to 11% of cases. Improved techniques have reduced deep infection rates to 1% to 2% [1]. Radiographic features may suggest infection, including signs of loosening, endosteal scalloping, and periosteal reaction. The sensitivity of endosteal scalloping, however, is only 47%, but the specificity is 96%. Periosteal new bone formation is only 25% sensitive but 92% specific [46]. Radionuclide scans, including PET imaging, and joint

aspiration with subtraction arthrography are more definitive [Figs. 16 and 17] [1]. Arthrograms may demonstrate irregular contrast extending around the components, irregular pocketing of the capsule, sinus tracts, or abscesses [see Figs. 16 and 17]. The authors aspirate the joint and send fluid for culture before contrast injection [1].

Osteolysis, as in the knee, may result from polyethylene wear or metal or cement reactivity. The incidence of this complication is approximately 5% [44]. Radiographs demonstrate lytic lesions about the components [Fig. 18]. CT, however, is used more often to evaluate the extent of bone loss

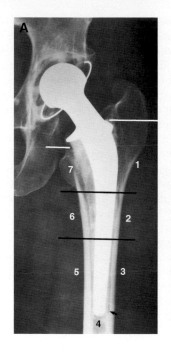

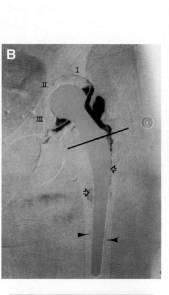

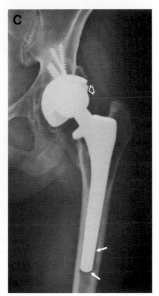

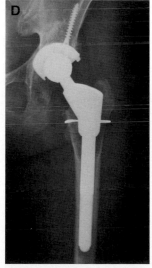

Fig. 15. Femoral component loosening. (*A*) The femoral component is divided into seven zones, with labeling from lateral to medial. There are lucent lines at the bone cement interface in this patient, most obviously at the tip (*arrow*) and in zones 3 and 5. (*B*) Subtraction arthrogram demonstrates irregular contrast about the upper femoral stem (*open arrows*) with more subtle changes distally (*arrowheads*) in zones 3 and 5 resulting from loosening. (*C*) Irregular lucency in zones 3 and 4 (*arrows*) about the uncemented acetabular and femoral components results from toggling. There is polyethylene wear with marked asymmetry of the femoral head (*open arrow*) in the acetabular component. (*D*) Both components were removed and a revision arthroplasty was performed.

fully before considering revision. Radiographs overlook up to 25% of lesions and the size of the lesions is evaluated more easily with CT. CT also is helpful for distinguishing osteolysis from geode formation [11,12].

Dislocations were discussed previously as a potential problem with initial weight bearing. There are many factors that increase the risk of dislocation [see Fig. 6], including operative approach, gender (females 8.9% and males 4.5% over 25 years),

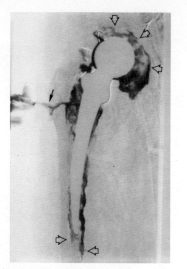

Fig. 16. Infected right hip prosthesis. Subtraction arthrogram shows extensive irregular contrast around both cemented components (*open arrows*) and a sinus tract (*arrow*) resulting from infection and loosening.

age (patients over 70 years age at the time of surgery dislocate more frequently than patients less than 70 years of age at the time of surgery), small femoral head size, component malposition, avulsion of the greater trochanter, and preoperative diagnosis [1,47–50]. Dislocations are detected easily on conventional radiographs [1].

Pseudobursae are identified in up to 43% of patients who have hip replacement. These are most common in the region of the greater and lesser trochanters and near the lateral acetabular margin. Pain is a common presenting complaint. Arthrography can diagnose the bursal locations and offers the opportunity for anesthetic injection to confirm the source of pain [51].

Greater trochanteric avulsion or nonunion after surgical osteotomy also may cause complications after arthroplasty. The trochanter usually is reattached with wire or a cable claw system if an osteotomy is performed. Union typically occurs in 12 weeks in 95% of patients. Unless there is significant (2-cm) displacement of an avulsed fragment or un-united osteotomy, reattachment may not be indicated. In the remainder of patients, union may be a problem and can have an impact on other potential complications, such as dislocation [1]. Radiographs are adequate for detection and evaluating fragment position.

Heterotopic ossification occurs to some degree after hip replacement in most patients. This usually is of little consequence. Problems are more likely in patients who have DISH; male patients who have osteoarthritis; and patients who have ankylosing spondylitis, previous surgery, or extensive surgical trauma [1,9]. Heterotopic ossification is graded based on the extent of involvement between the trochanter and acetabular join margin. Grade I involves 25% of the distance, grade II 50%,

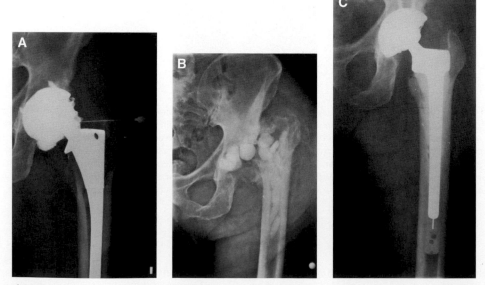

Fig. 17. Infected uncemented hip replacement. (A) Arthrogram demonstrates an irregular pseudocapsule. Aspiration isolated staphylococcus. (B) The implant was removed (Girdlestone procedure) and antibiotic beads were inserted. (C) After the infection was eradicated, a revision arthroplasty with a calcar femoral component and uncemented acetabular component was placed.

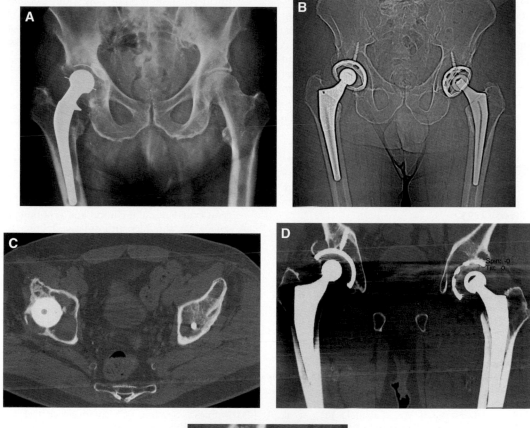

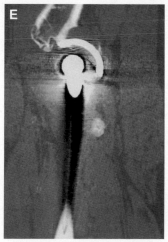

Fig. 18. Osteolysis. (*A*) AP radiograph of the pelvis demonstrates large irregular lytic areas about the right femoral component. (*B*) Scout CT image with bilateral implants and ceramic femoral head on the left in a different patient. CT images in the axial (*C*) plane, with coronal (*D*) and sagittal (*E*) reconstructions, more clearly demonstrate the extent of osteolysis. The multiple lucent foci with scalloped margins are classic for osteolysis resulting from particle disease.

and grade III 75%, and grade IV involves bone bridging the trochanter and acetabulum [1,9]. Extensive ossification can restrict motion and reduce function. Preoperative radiation may reduce the likelihood of extensive heterotopic ossification [9].

Summary

Radiologists play an important role in pre- and postoperative imaging of patients selected for joint replacement procedures. It is critical that the extent of bone loss, articular abnormalities, and soft tissue support are demonstrated before surgery. Patient and component selection are based on clinical findings and these important image features. Postoperatively, it is essential to understand what surgeons need to know regarding position and potential complications of a procedure. Optimal selection of imaging approaches is essential to defining the nature of potential complications, avoiding unnecessary revision surgery, or planning an appropriate surgical approach, if required.

References

[1] Berquist TH. Imaging atlas of orthopedic appliances and prostheses. New York: Raven Press; 1995.

[2] Mahomed NN, Barrett J, Katz JN, et al. Epidemiology of total knee replacement in the United States Medicare population. J Bone Joint Surg [Am] 2005;87-A:1222–8.

[3] Ethgen O, Bruyere O, Richy F, et al. Health-related quality of life in total hip and total knee arthroplasty. J Bone Joint Surg 2004;86-A:963–74.

[4] Anderson T, Montgomery F, Carlsson A. Uncemented STAR total ankle prostheses. J Bone Joint Surg [Am] 2003;85-A:1321–9.

[5] Rand JA, Trousdale RT, Ilstrup DM, et al. Factors affecting the durability of primary total knee prostheses. J Bone Joint Surg [Am] 2003;85-A:259–65.

[6] Berry DJ, Harmsen WS, Cabanela ME, et al. Twenty-five year survivorship of two thousand consecutive primary Charnley total hip replacements. Factors affecting survivorship of acetabular and femoral components. J Bone Joint Surg [Am] 2002;84-A:171–7.

[7] Kurtz S, Mowat F, Ong K, et al. Prevalence of primary and revision total hip and knee arthroplasty in the United States from 1990 through 2002. J Bone Joint Surg[Am] 2005;87-A:1487–97.

[8] Jones CP, Lachiewicz PF. Factors influencing the longer-term survival of uncemented acetabular components used in total hip revisions. J Bone Join Surg [Am] 2004;86-A:342–7.

[9] Rumi MN, Deol GS, Bergandi JA, et al. Optimal timing of preoperative radiation for prophylaxis against heterotopic ossification. A rabbit hip model. J Bone Joint Surg [Am] 2005;87-A:366–73.

[10] Udomkiat P, Wan Z, Dorr LD. Comparison of preoperative radiographs and intraoperative findings of fixation of hemispheric porous-coated sockets. J Bone Joint Surg [Am] 2001;83-A:1865–71.

[11] Kitamura N, Naudie DDR, Leung SB, et al. Diagnostic features of pelvic osteolysis on computed tomography: the importance of communication pathways. J Bone Joint Surg [Am] 2005;87-A:1542–50.

[12] Leung S, Naudie D, Kitamura N, et al. Computed tomography in assessment of periacetabular osteolysis. J Bone Joint Surg [Am] 2005;87-A:592–7.

[13] Stern SH, Insall JN. Posterior stabilized prosthesis. Results after follow-up of nine to 12 years. J Bone Joint Surg [Am] 1992;74-A:980–6.

[14] Collins DN, Heim SA, Nelson CL, et al. Porous coated anatomic total knee arthroplasty. Clin Orthop 1991;267:128–36.

[15] Patel DV, Ferris BD, Aichroth PM. Radiologic study of alignment after total knee replacement. Int Orthop 1991;15:209–10.

[16] Miura H, Matuda S, Mawatari T, et al. The oblique posterior femoral condylar radiographic view following total knee arthroplasty. J Bone Joint Surg [Am] 2004;86-A:47–51.

[17] Hammond JW, Queale WS, Kim TK, et al. Surgeon experience and clinical and economic outcomes for shoulder arthroplasty. J Bone Joint Surg [Am] 2003;85-A:2318–24.

[18] Parvizi J, Johnson BG, Rowland C, et al. Thirty-day mortality after elective total hip arthroplasty. J Bone Joint Surg [Am] 2001;83-A:1524–8.

[19] Della Valle CJ, Jazrawi LM, Idiadi J, et al. Anticoagulant treatment of thromboembolism with intravenous heparin therapy in the early postoperative period following total joint arthroplasty. J Bone Joint Surg [Am] 2000;82-A:207–12.

[20] Pellicci PM, Inglis AE, Salvati EA. Perforation of the femoral shaft during total hip replacement: report of 12 cases. J Bone Joint Surg [Am] 1980; 62-A:234–40.

[21] Kumar S, Sperling JW, Haidukewych GH, et al. Periprosthetic humeral fratures after shoulder arthroplasty. J Bone Joint Surg [Am] 2004;86-A:680–9.

[22] Berger RA, Meneghini M, Jacobs JJ, et al. Results of unicompartmental knee arthroplasty at a minimum of ten years of follow-up. J Bone Joint Surg. [Am] 2005;87-A:999–1006.

[23] Kitaoka HB, Patzer GL. Clinical results of Mayo total ankle arthroplasty. J Bone Joint Surg [Am] 1996;78-A:1658–64.

[24] Knecht SI, Estin M, Callaghan JJ, et al. The Agility total ankle arthroplasy. J Bone Joint Surg [Am] 2004;86-A:1161–71.

[25] Spirt AA, Assal M, Hansen ST. Complications and failure after total ankle arthroplasty. J Bone Joint Surg [Am] 2004;86-A:1172–9.

[26] Cracchiolo III A, Kitaoka HB, Leventen EO. Silicone implant arthroplasy for second metatarsophalangeal joint disorders with and without hallux valgus deformities. Foot Ankle 1988; 9:10–8.

[27] Cracchiolo III A, Weltmer JB, Lian G, et al. Arthroplasty of the first metatarsophalangeal joint. J Bone Joint Surg [Am] 1992;74-A:552–63.

[28] Bell SN, Geschwend N. Clinical experience with total arthroplasty and hemiarthroplasty of the shoulder using a Neer prosthesis. Int Orthop 1986;10:217–22.

[29] Werner CML, Steinmann PA, Gilbart M, et al. Treatment of painful pseudoparesis due to irrep-

arable rotator cuff dysfunction with the Delta III reverse-ball-and-socket total shoulder prosthesis. J Bone Joint Surg [Am] 2005;87-A: 1476–86.

[30] Ring D, Kocher M, Koris M, et al. Revision of unstable capitellocondylar (unlinked) total elbow replacement. J Bone Joint Surg [Am] 2005; 87-A:1075–9.

[31] Morrey BF. Complications of elbow replacement surgery. In: Morrey BF, editor. The elbow and its disorders. Philadelphia: W.B. Saunders; 2000. p. 667–77.

[32] Lee BP, Adams RA, Morrey BF. Polyethylene wear after total elbow arthroplasty. J Bone Joint Surg [Am] 2005;87-A:1080–7.

[33] Bechenbaugh RD. Arthroplasty of the wrist. In: Morrey BF, editor. Joint replacement arthroplasty. New York: Churchill-Livingstone; 1991. p. 195–215.

[34] Pakos EE, Ntzani EE, Trikalinos TA. Patellar resurfacing in total knee arthroplasty. J Bone Joint Surg [Am] 2005;87-A:1438–77.

[35] Pollock DC, Ammeen DJ, Ehgn GA. Synovial entrapment: a complication of posteriior stabilized total knee arthroplasty. J Bone Joint Surg [Am] 2002;84-A:2174–9.

[36] Gill ES, Mills DM. Long-term follow-up evaluation of 1,000 consecutive cemented total knee arthroplasties. Clin Orthop 1991;273:66–76.

[37] Dobbs RE, Haussen AD, Lewallen DG, et al. Quadriceps tendon rupture after total knee arthroplasty. J Bone Joint Surg [Am] 2005;87-A:37–45.

[38] Wood DJ, Smith AJ, Collopy D, et al. Patellar resurfacing in total knee arthroplasty. J Bone Joint Surg [Am] 2002;84-A:187–93.

[39] Conditt MA, Noble PC, Allen B, et al. Surface damage of patellar components used in total knee arthroplasty. J Bone Joint Surg [Am] 2005; 87-A:1265–71.

[40] Saleh KJ, Clark CR, Rand JA, et al. Modes of failure and preoperative evaluation. J Bone Joint Surg [Am] 2003;85-A(Suppl 1):21–5.

[41] Hoffmann AA, Wyatt RWB, Beck SW, et al. Cementless total knee arthroplasty in patients over 65 years of age. Clin Orthop 1991;271:28–34.

[42] Bengtson S, Knutson K. The infected knee arthroplasty: a six year follow-up on 357 cases. Acta Orthop Scand 1991;62:301–11.

[43] Tsukayama DT, Goldberg VM, Kyle R. Diagnosis and management of infection after total knee arthroplasty. J Bone Joint Surg [Am] 2003; 85-A(Suppl 1):75–80.

[44] Park Y-S, Moon Y-W, Lim S-J, et al. Early osteolysis following second-generation metal-on-metal hip replacement. J Bone Joint Surg [Am] 2005; 87-A:1515–21.

[45] Urban JA, Garvin KL, Boese CK, et al. Modes of failure and preoperative evaluation. J Bone Joint Surg [Am] 2001;83-A:1688–94.

[46] Lyons CW, Berquist TH, Lyons JC, et al. Evaluation of radiographic findings in painful hip arthroplasties. Clin Orthop 1985;195:239–51.

[47] Berry DJ, von Knoch M, Schleck CD, et al. The cumulative long-term risk of dislocation after primary Charnley total hip replacement. J Bone Joint Surg [Am] 2004;86-A:9–14.

[48] Hermida JC, Bergula A, Chen P, et al. Comparison of wear rates of twenty-eight and thirty-two millimeter femoral heads on cross-linked polyethylene acetabular cups in a wear simulator. J Bone Joint Surg [Am] 2003;85-A:2325–31.

[49] Bartz RL, Noble PC, Kadakia NR, et al. The effect of femoral component head size on posterior dislocation of the artificial hip joint. J Bone Joint Surg [Am] 2000;82-A:1300–7.

[50] Goldstein WM, Gleason TF, Kopplin M, et al. Prevalence of dislocation after total hip arthroplasty through a posterolateral approach with partial capsulotomy and capsulorrhaphy. J Bone Joint Surg [Am] 2001;83-A:2–7.

[51] Berquist TH, Bender CE, Maus TP, et al. Pseudobursae: a useful finding in patients with painful hip arthroplasty. AJR Am J Roentgenol 1987; 148:103–6.

RADIOLOGIC
CLINICS
OF NORTH AMERICA

Radiol Clin N Am 44 (2006) 439–450

Postoperative Infection

Jeffrey J. Peterson, MD

Infectious disease complicating prior musculoskeletal surgery continues to be one of the most important causes of postoperative morbidity and mortality. Postoperative infection often presents with nonspecific pain and swelling and can be difficult to diagnose accurately. Timely detection and accurate localization of infectious processes have important clinical implications and are critical to appropriate patient management. Discrimination between infectious and noninfectious inflammation continues to pose a diagnostic challenge, and the search continues for methods that can reveal infectious foci rapidly and effectively. This article reviews the various methods and modalities available for the detection of postoperative infection. It is not intended to be a comprehensive review but a review of the basic concepts and fundamental principles of imaging in postsurgical infection.

The goal of surgical intervention in orthopedics generally is to alleviate pain and restore function. A small percentage of cases are complicated, however, by postoperative infection, despite the fact that surgical techniques and antibiotic therapy continue to improve. There are many variables that contribute to postoperative infection. These can be divided into procedural variables (including type and length of procedure) and patient variables (such as the general medical and physical condition of patients before surgery). Probably the most basic, yet critical, variable is the type of surgery performed [1]. Simple procedures with short operative times and minimal incisions typically result in lower rates of postoperative infection than more complex procedures with long operative times and large incisions [1]. Procedures requiring prolonged operative time (>3 hours) result in higher postoperative infections rates compared with shorter procedures [2]. High blood loss (>1000 mL) also poses an increased risk for infection [2]. Operative procedures resulting in surgical instrumentation or hardware are shown to increase the risk of infection by between 3% and 6% [1]. The overall incidence of infections related to orthopedic devices is variable, ranging between 0.5% and 2% [3]. Procedures requiring multiple incisions and operative sites, such as procedures requiring additional bone graft donor sites, increase the risk of infection. It is estimated that the complication rate of the separate bone graft donor site alone is up to 20% [1]. Staged operative procedures with more than one surgery performed on differing days also increases infection rates compared with same-day procedures [1,4].

Patient factors play a role in infection rates. Generally, younger and healthier patients are more resistant to infection compared with older, debilitated patients. Studies show that infection rates in patients younger than 20 years old statistically are

Department of Radiology, Mayo Clinic, 4500 San Pablo Road, Jacksonville, FL 32224-3899, USA
E-mail address: Peterson.Jeffrey@mayo.edu

0033-8389/06/$ – see front matter © 2006 Elsevier Inc. All rights reserved.
radiologic.theclinics.com
doi:10.1016/j.rcl.2006.01.007

lower than in patients older than 20 years of age [1,5]. Obesity and tobacco use are linked to higher risk of postoperative infection. Preoperative illness and medical disorders factor into the risk of postoperative infection. Extended periods of prehospitalization and malnutrition prove to result in higher infection rates after surgery [1]. This also is seen in staged operative procedures, where alterations in nutritional status after the primary procedure is believed to contribute to the higher rate of complications for secondary staged procedures [4].

Surgical procedures allow direct inoculation of the deep tissues with microorganisms. The fundamental step in starting an infection is bacterial colonization [6]. Healthy soft tissue and bone are extremely resistant to infection under normal circumstances. After the tissue trauma inherent in surgical procedures, however, the resistance of the affected tissue to infection decreases substantially. Instrumentation and the addition of hardware and other foreign bodies into the tissues provide scaffolding conducive to infection. Interruption of the blood supply during surgery and resultant compromise of the microcirculation after surgery can result in devitalized tissue, which is further susceptible to infection. These devitalized surfaces, in conjunction with instrumentation and local hematoma, offer an ideal culture medium for bacterial growth [7]. Activated cytokines released by tissues traumatized by surgical intervention also have a deleterious effect on the host response to infection [8].

The most common organism found in postoperative infections is staphylococcus. With peripheral joint arthroplasty, *Staphylococcus epidermidis* is the most common offending agent (31%) followed by *Staphylococcus aureus* (20%) [9]. In spinal surgery, it is reported that *Staphylococcus aureus* is the most common organism isolated, accounting for 49% of cases, followed by *Staphylococcus epidermidis* in 28% [1]. Single organism infection is more common than multiple organisms, with only 8.3% of cases of postoperative infection containing gram-positive and gram-negative organisms [1]. To reduce the growth of the more common pathogens found in postoperative infection, preoperative prophylaxis with first-generation cephalosporins may be used [10].

Diagnosis

In the immediate postoperative period, complaints of pain and discomfort frequently are encountered. Most of the time, the pain is attributed to the trauma related to the procedure. When symptoms increase after a period of comfort, however, postoperative infection may have developed [1]. Inflammatory changes after surgery normally begin

to subside 5 to 7 days after the operation. During this time, the signs and symptoms of acute infection often become gradually evident [11]. Pain is a universal symptom; however, fever often is absent. External signs of infection (erythema, warmth, and edema), although nonspecific, can be important clues to underlying infection. The operative wound rarely has a benign appearance at presentation and wound drainage is seen in 93% of cases [12]. With joint arthroplasty, it is estimated that one third of patients who have postoperative infection develop their infection within 3 months, and another one third develop them within 1 year [13]. Chronic infection after orthopedic surgery frequently is more indolent than acute infection and can be more difficult to diagnose. Often the symptoms are less severe and less specific and may mimic other musculoskeletal disorders. In patients who have orthopedic hardware or joint arthroplasty, chronic infection can be difficult to differentiate from mechanical or hardware failure. Differentiation of aseptic loosening of orthopedic hardware from infection is challenging, clinically and histopathologically.

Laboratory analysis may be useful in the detection of postoperative infection. Elevated leukocyte count often is seen; however, this is neither sensitive nor specific. Elevated values can be seen normally in the postoperative period and, therefore, prove unreliable as an indicator of infection in the immediate postoperative period. C-reactive protein and erythrocyte sedimentation rates may be elevated; however, these values are nonspecific and may be elevated in the uncomplicated postoperative period. A study of patients after uncomplicated spine surgery finds that C-reactive protein values typically peak 2 to 3 days after the procedure and return to normal within 5 to 14 days [2]. Erythrocyte sedimentation rate peaks at 5 days after surgery but has a more variable decline to normal, often remaining elevated for 21 to 42 days [2].

Culture is the most definitive analysis for postoperative infection, although this not always is accessible after surgery. Wound cultures obtained from postoperative defects or open sinus tracts can be useful for diagnosis of infection and isolation of the offending organism, which can be targeted by specific antibiotic therapy. Aspiration of affected joints or associated fluid collections also can be useful in the appropriate setting.

Although often lacking in sensitivity, culture is highly specific when positive [14]. Postoperative infection often is associated with bacteremia and blood cultures may be of usefulness in some cases. Culture of the surgical wound or draining sinus tract often is successful at isolating organisms, although cultures should be taken from the deepest point of the wound or tract to avoid contaminants from the surrounding skin surfaces. Aspiration of fluid collections, often under image guidance, provides material for culture and can aid in differentiating abscess from bland postoperative fluid collections (hematoma or seroma). Aspiration of affected joint cavities, with or without image guidance, may be useful; however, sensitivity of joint aspiration for the detection of infection often is suboptimal. In a prior study of 33 patients who had suspected joint infection after joint arthroplasty, 19 of whom had infection based on intraoperative findings, aspiration cultures were 74% sensitive [15]. Cuckler and colleagues reviewed nine prior studies retrospectively and found the overall sensitivity of joint aspiration for the detection of joint infection was only 68% [16]. Biopsy of soft tissue or bone may be used, although these techniques yield a low positive predictive value in the accurate identification of the causative microorganism [17]. Biopsy inherently risks introducing infection into a potentially uninfected field.

Imaging

Imaging studies can play an important role in the detection of infection and can help guide appropriate clinical management. The diagnosis of postoperative infection can be made with a variety of imaging modalities. Radiographs often are the initial technique used for the evaluation of suspected postoperative infection and can be helpful to exclude other cases of postoperative pain. Cross-sectional anatomic imaging studies, such as CT and MR imaging, can be useful for more detailed depiction of the affected area and can further delineate associated soft tissue and bony abnormalities seen on radiographs. Unfortunately, metallic hardware after surgery may degrade image quality with CT and MR imaging. In cases where cross-sectional imaging quality is degraded significantly by metal or in patients in whom MR imaging is contraindicated, nuclear medicine imaging, such as bone scintigraphy or radiolabeled leukocyte scans, can detect and localize sites of infection reliably in the postoperative period.

Conventional radiographs

The mainstay of imaging in the postoperative period is conventional radiography. Radiographs frequently are the first imaging study obtained in patients who have suspected postoperative infection. Although radiographic findings may lag behind clinical disease, radiographs can identify soft tissue swelling and loss of soft tissue planes within 3 days of bacterial contamination [18]. Post-

operative osseous infection also can be seen on radiographs as focal areas of cortical erosion, destruction, or osteolysis [Fig. 1]. With acute bacterial infections, bone lysis may be detected 7 to 14 days after infection [18]. Infectious involvement of the periosteum can stimulate periosteal reaction, which also can be detected radiographically. Chronic infections often produce eburnation and sclerosis within the bone. Radiographs can diagnose other disorders in postoperative patients and help differentiate infection from other abnormalities, such as fracture or hardware failure. In recently postoperative patients, radiographs can be used to exclude retained foreign bodies within the surgical wound.

Detection of infectious processes within the bone can be difficult in osseous structures that are altered by previous surgical procedures. Cortical irregularity, periosteal reaction, osteolysis, and sclerosis all can be seen postoperatively on radiographs, even in the absence of infection. The sensitivity and specificity for infection with endosteal scalloping seen on radiographs is 47% and 96%, respectively, and 25% and 92% for periosteal reaction [19]. Arthroplasty loosening or polyethylene osteolyis may present with radiographic appearances similar to infection [14]. The radiographic findings of soft tissue infection also are nonspecific and can be seen in the absence of infection in the immediate postoperative period. Edema or hemorrhage may

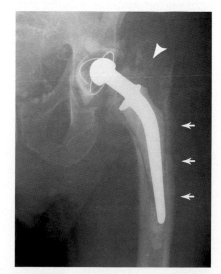

Fig. 1. Conventional radiograph of an infected left hip arthroplasty in a patient who had chronic left hip discomfort. Lucency is seen at the bone-cement interface of the acetabular and femoral components, compatible with gross loosening. Osteolysis also is seen in the left greater trochanteric region (*arrowhead*). Extensive irregular periosteal reaction about the left proximal femur (*arrows*) strongly suggests infection, which was confirmed with aspiration.

simulate cellulitis radiographically. In such cases, cross-sectional imaging with CT or MR may be useful. Temporal evaluation of radiographs is useful when infection is suspected. It is important to compare current studies with prior radiographs, when available, to evaluate for changes with time. Development of otherwise nonspecific radiographic abnormalities, such as osteolysis or cortical destruction, in a short period of time should be viewed as worrisome for infection.

Arthrography is suggested as useful for evaluation of joint infection in patients who have prior arthroplasty or other joint intervention. Findings for postoperative infection include irregular, enlarged pseudocapsules or sinus tracts, with or without cavities [20]. Percutaneous aspiration of joint fluid may be useful and can be performed with or without image guidance. Aspiration can yield a specific organism and guide antibiotic treatment; sensitivity for joint aspiration, however, is varied in the literature and generally regarded as less than optimal [14].

Despite its limitations, radiographs remain the best initial examination in cases of symptomatic postoperative patients. They are quick and inexpensive and may depict findings associated with postoperative infection. Given the lack of distortion from metallic implants or hardware, radiographs aid in the selection and interpretation of more advanced, second-line imaging modalities, such as CT, MR, or nuclear medicine imaging studies [3].

MR imaging

MR imaging has proved a useful imaging modality for the detection of infectious processes. MR imaging can identify soft tissue inflammatory changes reliably and osseous abnormalities associated with postoperative infection [3]. The primary criteria for osseous and soft tissue infection is increased signal intensity on T2-weighted images or other fluid sensitive sequences, decreased signal intensity on T1-weighted images, and the presence of contrast enhancement after the administration of intravenous gadolinium (related to increased vascular permeability) [Fig. 2] [3].

MR imaging can depict soft tissue inflammatory processes accurately, such as cellulitis and phlegmonous inflammation of the soft tissues and associated postoperative fluid collections or abscess formation. Fistulas or sinus tracts can be demonstrated precisely with MR imaging examinations. MR imaging can help differentiate abscess from postoperative hematoma. Areas of prior hemorrhage typically depict high signal on T1-weighted images or low signal on fluid sensitive sequences, depending on the age of the hematoma. Gadolinium administration can aid in differentiating abscess from bland fluid collections, such as seromas. Typically, the wall of an abscess cavity is thick and irregular and enhances profusely; the margins of a bland postoperative fluid collection typically are thin and demonstrate little enhancement.

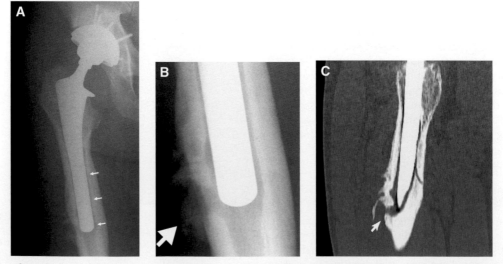

Fig. 2. Infected hip prosthesis in an older gentleman 2 years after implant placement. (A) Radiograph depicts a noncemented right total hip arthroplasty with lucency about the femoral component (arrows). (B) There is irregular periosteal reaction lateral to the tip of the femoral component and lucency (arrow) traversing the lateral femoral cortex, consistent with a sinus tract. (C) CT confirms osteolysis about the femoral component and sinus tract extending from the distal tip of the component into the lateral soft tissues (arrow).

Gadolinium administration can aid in differentiation between chronic infection and fibrovascular scar. Less enhancement is expected in areas of fibrovascular scar compared with areas of infection.

In addition to soft tissue infection, MR imaging reliably can demonstrate the presence of osseous infection in postoperative patients. Osseous infection typically results from contiguous spread of soft tissue infection; however, when prior surgery involves the bone, direct inoculation of the bone can occur. MR imaging typically depicts osseous infection as areas of edema-like signal within the osseous structure, with low signal intensity on T1-weighted images and high signal on fluid-sensitive sequences [21]. Areas of cortical destruction can be seen with loss of the low signal intensity cortical margins [21]. Early periosteal involvement can be seen as linear high signal on T2-weighted images or fluid sensitive sequences extending along the periosteal surface. Osseous infection often is contiguous with the previous operative site or adjacent to sites of soft tissue infection.

MR imaging has some limitations in postoperative patients. Although areas of abnormal signal can identify sites of infection readily, similar patterns are seen in noninfected, postoperative patients at sites of soft tissue trauma and inflammation attributed to the procedure itself. This abnormal signal related to surgery can last for long periods of time, limiting the usefulness of MR imaging. In addition, MR imaging may be hindered significantly by metal introduced at surgery [3]. This can cause gross misregistration and field distortion with large pieces of metal and foci of rounded susceptibility artifact with smaller metallic fragments. Fat saturation or water excitation techniques are hindered by the presence of metal.

Before obtaining MR imaging, however, radiographs should be performed, as some findings (soft tissue calcification and periosteal reaction) may be visualized better on radiographs. In cases with significant metallic hardware, nuclear medicine imaging techniques may be better suited for evaluation of postoperative infection.

CT

CT has become useful in the detection of postoperative infection and can be a helpful adjunct to radiographs. Modern multislice CT can obtain thinly collimated images quickly that can be reformatted into multiplanar reconstructions or 3-D volume rendered or surface-shaded images. CT depicts the osseous structures precisely and allows visualization of subtle erosion and periosteal reaction. CT can precisely depict bony sequestra, intraosseous fistulae, cortical tracts, soft tissue defects,

and sinus tracts [Fig. 3] [3]. Soft tissue infection can be seen with CT; however, the soft tissue contrast is less with CT than with MR. Iodinated contrast media can be helpful in the evaluation of enhancement patterns in cases of suspected postoperative soft tissue infection. Infectious foci tend to enhance significantly; thick, irregular rim enhancement can be seen with abscess formation. In addition to diagnostic imaging, CT can be useful in guiding interventions (aspirations or drain placement) in areas of soft tissue infection or bone biopsy in cases of suspected osteomyelitis.

The soft tissue contrast resolution is suboptimal for CT and subtle soft tissue infection may be difficult to appreciate. Specificity for soft tissue infection can be problematic, as infectious foci may be difficult to differentiate from other postoperative change, such as edema and hematoma. In addition, CT is adversely affected by metal hardware [10]. CT in patients who have metal implants often is hindered by streak artifact caused by inability of the x-ray beam to penetrate dense metal, which has a high attenuation coefficient. This results in image reconstructions with missing projection data. Artifacts are seen as starburst streaking radiating from the metallic component, which often severely degrades image quality and hinders the evaluation of the surrounding osseous structures and soft tissues.

Nuclear medicine imaging

Although helpful in many instances, cross-sectional imaging often is degraded by image distortion and artifacts related to metallic prostheses or hardware placed at surgery. This can result in findings that are nonspecific or unclear. In such cases, nuclear medicine investigations provide an additional modality to evaluate for the presence of infection in postoperative patients [22]. Nuclear medicine examinations provide important physiologic information that, in some cases, may precede anatomic abnormalities, allowing for more timely detection and treatment of postoperative infection.

Currently, four radionuclides are used commonly in the United States for the detection of infection: technetium TC 99m (^{99m}Tc) methylene diphosphonate (MDP), indium In 111 (^{111}In)–labeled leukocytes, ^{99m}Tc-labeled leukocytes, and gallium citrate Ga 67 citrate (^{67}Ga). In recent years, F18-fluorodeoxyglucose positron emission tomography (^{18}F-FDG PET) has emerged as an alternative to traditional scintigraphic imaging and has gained interest for its potential role in detecting infectious and inflammation. Several additional agents, such as ^{99m}Tc fanolesomab (NeutroSpec, Mallinckrodt Imaging, St. Louis, Missouri, an antigranulo-

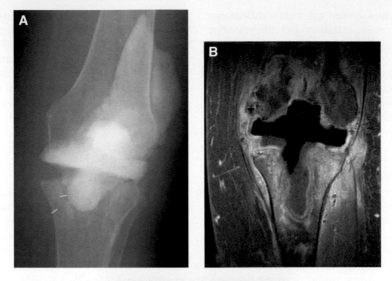

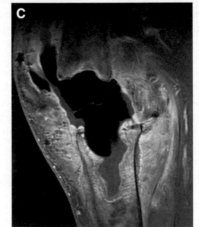

Fig. 3. Status post resection of infected knee arthroplasty with antibiotic spacer placement. (*A*) Radiograph depicts postoperative changes with an antibiotic laden cement spacer in the knee joint. (*B, C*) Post gadolinium T1-weighted MR images depict residual inflammatory change and fluid within the knee joint and the surrounding soft tissues. Persistent osseous infection is depicted by enhancement throughout the proximal tibia, fluid and inflammatory changes within and about the tibial component defect.

cyte antibody, currently are under investigation but have yet to be in widespread clinical use.

Technetium Tc 99m bone scintigraphy

In today's practice, ^{99m}Tc-labeled compounds account for approximately 80% of all radiopharmaceuticals used in clinical nuclear medicine. This can be explained by its favorable imaging characteristics, widespread availability, and low cost. The two technetium preparations used most commonly for the detection and evaluation of infection are bone scintigraphy and ^{99m}Tc-labeled leukocytes.

The radiopharmaceuticals used most commonly for bone scintigraphy are ^{99m}Tc phosphonates, spe-

cifically, ^{99m}Tc-MDP. Bone scintigraphy is available widely, performed easily, and can be useful in the detection of postoperative infection involving the osseous structures. With bone scintigraphy, the radiopharmaceutical localizes to the bone surface by a process known as chemiabsorption. The radiotracer is absorbed onto osseous surfaces, with its uptake depending on blood flow and osteoblastic activity. Single photon emission CT (SPECT) imaging can be performed, allowing multiplanar imaging and improving lesion detection and localization. Bone scintigraphy is exquisitely sensitive; lack of accumulation of radiopharmaceutical at a site of suspected osseous infection is strong evidence against osteomyelitis. Unfortunately, bone scintigraphy demonstrates suboptimal specificity and the significance of uptake at a site of suspected infection is less certain [9]. Specificity of bone scintigraphy for infection is estimated to be as low as 18% to 25% [3]. Several differing disorders (benign and malignant) may result in increased osseous scintigraphic activity, including trauma, infection, arthropathy, and neoplasm.

Three-phase bone scan can be performed to improve the accuracy of standard bone scintigraphy in the detection of postoperative infection. The three phases include dynamic blood flow imaging, a blood pool image, and delayed imaging [Fig. 4]. Blood flow and blood pool images constitute the early phase of imaging and demonstrate the presence and distribution of hyperemia. In addition to osseous involvement, early phase imaging may demonstrate soft tissue infectious processes that may not be depicted on routine imaging. Accuracy of three-phase bone scintigraphy for the detection of osteomyelitis is reported to approach 90% [23].

Specificity of bone scintigraphy in postoperative patients is suboptimal. Differentiation between infection and aseptic loosening after joint arthroplasty can be difficult with bone scanning. Particular patterns of activity are described to help differentiate infectious versus noninfectious uptake about specific joint prosthesis. The type of prosthesis (cemented or porous-coated non-cemented) and date of prosthetic placement must be known, however. In addition, some institutions report that up to two thirds of prostheses become infected within 1 year of surgery. During this period, uptake is so variable that only a normal bone scan yields diagnostic information [5]. Overall accuracy of radionuclide imaging in the evaluation of prosthetic joints is approximately 50% to 70%; however, it continues to be useful as an initial examination because of its high negative predictive value [24].

Radiolabeled leukocyte imaging

Radiolabeled leukocytes often are used to evaluate for postoperative infection. Although sensitivity is on par with conventional or three-phase bone scintigraphy, specificity increases significantly with radiolabeled leukocytes compared with bone scintigraphy. The two radiolabeled leukocyte examinations used commonly are ^{111}In-labeled leukocytes and ^{99m}Tc–hexamethylpropyleneamine oxime (HMPAO)–labeled leukocytes. Typically, neutrophils predominantly are labeled and represent the active component of a labeled leukocyte preparation. Lymphocytes are radiosensitive and typically fail to recirculate after the radiation exposure encountered in the labeling process. Given the predominant neutrophil tag, radiolabeled leukocyte imaging studies are most useful in detecting neutrophil-mediated inflammatory responses in acute postoperative infections. They are less useful in evaluating inflammatory processes involving a cel-

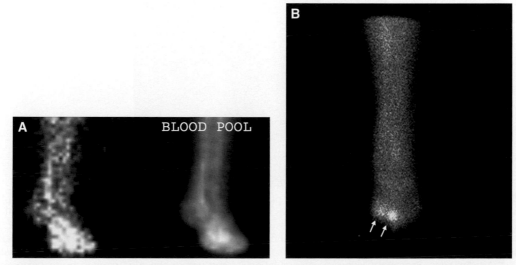

Fig. 4. A 58-year-old woman status post amputation of the great toe for osteomyelitis, who had recurrent signs of infection in the medial forefoot. (*A*) Dynamic images from a three-phase bone scan depict hyperemia throughout the left forefoot seen on both the dynamic blood flow images (*left*) and the blood pool image (*right*). (*B*) Routine 4-hour image depicts increased radionuclide accumulation in the left first and second toes (*arrows*). Findings were compatible with recurrent osteomyelitis involving the left first digit and development of osteomyelitis in the second digit and confirmed surgically.

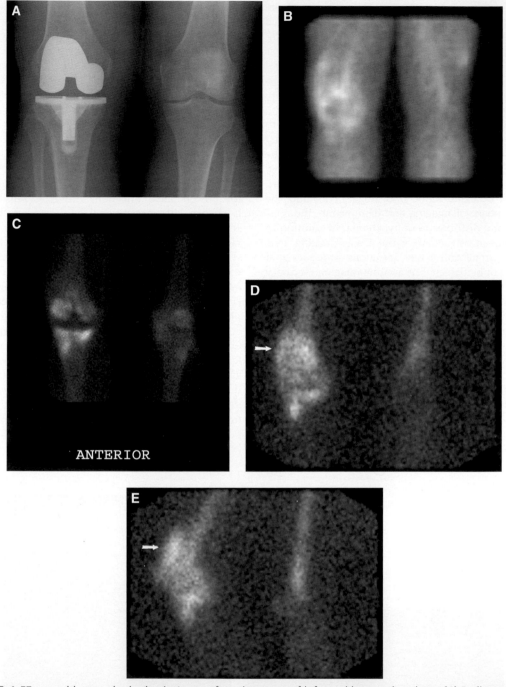

Fig. 5. A 57-year-old man who had pain 1 year after placement of left total knee arthroplasty. (*A*) Radiograph of the arthroplasty is unrevealing. (*B*) Blood pool images from a three-phase bone scan depict hyperemia about the right knee. (*C*) Four-hour bone scan image reveals a photopenic defect related to the metallic implant and nonspecific, increased radionuclide accumulation about the right knee joint. (*D, E*) Anterior (*D*) and LAO (*E*) images from an ^{111}In-labeled leukocyte scan confirm infection of the implant, with marked leukocyte localization about the right knee. Prominent leukocyte accumulation is seen within the suprapatellar bursa (*arrows*). At surgery, the suprapatellar bursa was distended with purulent material.

lular response other than neutrophils, such as chronic infection [25].

Radiolabeled leukocyte scans commonly are used for the evaluation of postoperative infection. Investigations of the efficacy of leukocyte scans vary widely in the literature, with sensitivities ranging from 50% to 100% and specificities from 45% to 100% [26]. Reported low sensitivity is attributed to several factors; most prominent is chronicity of the joint infections. Not uncommonly, chronic periprosthetic infections are low grade and elicit little in the way of inflammation.

[111]In-labeled leukocytes currently are considered the gold-standard nuclear medicine technique for the detection of postoperative infection [Fig. 5]. Leukocytes can be labeled effectively with ^{111}In, which has fair imaging characteristics. After intravenous administration, radiolabeled leukocytes accumulate rapidly at predominantly neutrophilic infiltrates [27]. Various studies show the efficacy of ^{111}In-labeled leukocytes for the detection of infectious foci, with sensitivity values ranging from 84% to 96% and specificity greater than 96% [28]. The advantage of ^{111}In-labeled leukocytes over ^{99m}Tc labeled leukocytes is that the preparation is more stable over time; therefore, background activity is less of a problem. The long half-life allows imaging of cell migration over 48 hours: this improves detection of low-grade chronic postoperative infections, which often demonstrate reduced migration of leukocytes [3]. A disadvantage of ^{111}In-containing preparations is the complex and time-consuming labeling process that requires handling of blood products. Indium is not available as readily as technetium, has higher radiation exposures, and has inferior imaging characteristics compared with technetium preparations.

Labeling of leukocytes with ^{99m}Tc-HMPAO has several advantages compared with Indium [Fig. 6]. Advantages of technetium preparations include continuous availability, lower radiation exposures, rapid examination, and better imaging characteristics. ^{99m}Tc-labeled leukocytes are less stable compared with ^{111}In preparations, and approximately 5% to 7% of the radiolabeled elutes from the white blood cells every hour. This leads to unwanted background activity that can obscure subtle sites of leukocyte accumulation.

Gallium citrate Ga 67

The first nuclear medicine imaging agent used for imaging of inflammation was gallium. To this day, it still is not understood precisely how it localizes. It is known that ^{67}Ga binds avidly to transferrin in the bloodstream, forming a gallium-transferrin macromolecule that localizes at sites of infection, via extravasation related to increased endothelial permeability. It is believed that at the site of infection, gallium disassociates from tranferrin and forms a gallium-lactoferrin complex [24]. Imaging typically is performed 48 to 72 hours after intravenous injection, and SPECT imaging can be obtained to improve on routinely obtained planar imaging. Overall accuracy for gallium in detecting foci of inflammation ranges from 70% to 80%.

Gallium is accumulated in a wide range of inflammatory, infective, and neoplastic processes

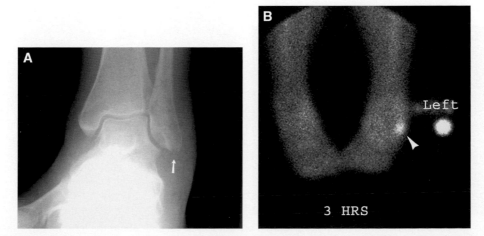

Fig. 6. A 45-year-old female who had continued drainage from the surgical wound after peroneal tendon repair. (*A*) Radiograph depicts soft tissue swelling overlying the lateral malleolus and subtle focal osteopenia of the lateral malleolus (*arrow*). (*B*) ^{99m}Tc-labeled leukocyte scan depicts increased leukocyte accumulation in the distal left fibula (*arrowhead*), compatible with osteomyelitis. Findings were confirmed at surgery and the region was debrided.

but is hindered by unfavorable imaging characteristics: it emits high-energy gamma rays that decrease spatial resolution. Gallium studies have a long examination time, requiring up to 7 days to image, with resultant high radiation exposure to the patient [3]. Because of the unfavorable imaging characteristics, gallium largely is replaced by other radiopharmaceuticals, such as radiolabeled leukocyte imaging and PET.

F18-fluorodeoxyglucose positron emission tomography

One of the potential areas of usefulness for PET currently studied is the detection of infection. Preliminary investigations of the use of FDG PET for the detection of musculoskeletal infection are promising [Fig. 7]. DeWinter and colleagues find FDG PET alone to be more accurate than the combination of bone-scan and white blood cell scan for the diagnosis of musculoskeletal infection in a series of 34 patients. Accuracy of PET was 94% compared with 81% for the combination of bone scan and leukocyte scan [22].

The radionuclide used most commonly for PET is FDG. FDG behaves like glucose in vivo and provides a means of quantitating the glucose metabolism. Areas of active inflammation have higher rates of glycolysis than uninvolved areas, relating primarily to the high concentrations of activated macrophages and leukocytes within the inflammatory response.

PET demonstrates several advantages over other conventional nuclear medicine techniques. Imaging and interpretation can be performed rapidly with results available in as little as 2 hours. PET demonstrates higher resolution compared with other nuclear medicine techniques and provides multiplanar imaging. An additional advantage is low uptake in bone marrow. PET, therefore, can differentiate between hematopoetic marrow and activated white blood cells, has higher accuracy in the central skeleton, and has increased sensitivity for low-grade infections. One potential drawback is the lack of anatomic landmarks; however, this can be overcome with comparison to anatomic imaging. Many scanners now obtain CT scans with the PET scan in one machine.

Much interest exists in the potential use of FDG PET in the delineation of infected prostheses versus mechanical loosening or polyethylene osteolysis. Preliminary results demonstrate FDG PET to be highly sensitive for the detection of infected prostheses; however, specificity is a concern and a significant problem with false-positive examinations is reported [29]. Histopathologic studies performed on tissue samples for revision arthroplasties for painful hip prostheses reveal that inflammatory cells accumulate in the periprosthetic soft tissue in infected and noninfected prostheses [30]. It seems that macrophage activity seen in cases of polyethylene osteolysis also use glucose, resulting in increased FDG accumulation about the joint in noninfected prostheses. Location of uptake about hip prostheses is suggested as useful in delineating infected prostheses. Activity along the shaft or distal femoral prosthesis is suggestive of infection, whereas uptake about the head and neck is considered nonspecific. The authors' anecdotal experience agrees with the lack of specificity of uptake about the head and neck for infection. Noninfectious activity can be seen along the femoral shaft

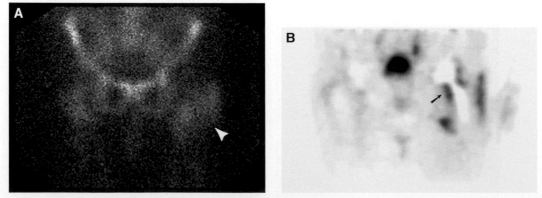

Fig. 7. An 85-year-old patient who had painful left hip prosthesis and cellulitis about the left hip. (*A*) ^{111}In-labeled leukocyte scan depicts significant leukocyte localization about the left hip (*arrowhead*), compatible with infection. It was difficult, however, to confidently define abnormal intraarticular activity secondary to the degree of activity seen in the surrounding soft tissues. (*B*) Multiplanar capability of [18]F- FDG PET allows clear delineation of increased metabolic activity in the soft tissues surrounding the hip and the hip joint itself (*arrow*). This required resection of the hip arthroplasty, which was confirmed to be infected.

and distal femoral component. Atypical cases of polyethylene osteolysis can result in significant FDG uptake anywhere along the components.

Investigational agents

Several new agents currently are being studied for potential uses in the detection of infectious processes. Many involve the use of antibodies to aid in more selective discrimination of targets and attempt to increase the sensitivity. A promising new agent is ^{99m}Tc fanolesomab (NeutroSpec), which is a ^{99m}Tc anti-CD15 IgM murine monoclonal antibody shown to have high affinity of CD15 receptors expressed on human neutrophils [31]. The antibody solution tagged with ^{99m}Tc is injected intravenously and the antibodies label the neutrophils rapidly. Imaging then is obtained to evaluate for sites of inflammation, indicated by neutrophil aggregation. Imaging can be performed rapidly after administration of the antibody solution and no handling of blood products is necessary. Preliminary results are promising and several centers are beginning to use NeutroSpec for clinical applications.

Summary

Although operative techniques and antibiotic therapy continue to improve, infection is an important cause of morbidity in postoperative patients. Accurate and timely diagnosis of postoperative infection is critical in appropriate management of these patients. Detection and localization of postoperative infection can be a challenging task. Physical examination and laboratory evaluation play important roles but often are hindered significantly in patients who have recent surgery. Imaging studies, therefore, are used commonly to supplement the search and evaluation for infectious foci. Radiographs have proved helpful as an initial examination, whereas cross-sectional imaging, such as CT or MR, often is more accurate in their ability to demonstrate soft tissue and osseous infection. In cases where cross-sectional images are compromised by metallic hardware, nuclear medicine examinations can identify foci of postoperative infection accurately. Imaging techniques continue to evolve and improve, and the search for better ways to image and detect postoperative infection will continue into the future.

References

[1] Beiner JM, Grauer J, Kwon BK, et al. Postoperative wound infections of the spine. Neurosurg Focus 2003;15:1–5.

[2] Thelander U, Larsson S. Quantitation of C-reactive protein levels and erythrocyte sedimentation rate after spinal surgery. Spine 1992;17:400–4.

[3] Kaim AH, Gross T, von Schulthess GK. Imaging of chronic posttraumatic osteomyelitis. Eur Radiol 2002;12:1193–202.

[4] Dick J, Boachie-Adjei O, Wilson M. One-stage versus two-stage anterior and posterior spinal reconstruction in adults. Comparison of outcomes including nutritional status, complication rates, hospital costs, and other factors. Spine 1992; 17(8 Suppl):S310–6.

[5] Capen DA, Calderone RR, Green A. Perioperative risk factors for wound infections after lower back fusions. Orthop Clin North Am 1996;27:83–6.

[6] Dirschl DR, Almekinders LC. Osteomyelitis. Common causes and treatment recommendations. Drugs 1993;45:29–43.

[7] Roesgen M, Hierholzer G, Hax P. Post-traumatic osteomyelitis. Arch Orthop Trauma Surg 1989; 108:1–9.

[8] Tsukayama DT. Pathophysiology of posttraumatic osteomyelitis. Clin Orthop 1999;360:22–9.

[9] Love C, Tomas MB, Marwin SE, et al. Role of nuclear medicine in diagnosis of the infected joint replacement. Radiographics 2001;21:1229–38.

[10] Horwitz NH, Curtin JA. Prophylactic antibiotics and wound infections following laminectomy for lumbar disc herniation. J Neurosurg 1975;43: 727–31.

[11] Ehara S. Complications of skeletal trauma. Radiol Clin North Am 1999;35:767–81.

[12] Weinstein MA, McCabe JP, Camissa FP. Postoperative spinal wound infection: a review of 2391 consecutive index procedures. J Spinal Disord 2000;13:422–6.

[13] Tsukayama DT, Estrada R, Gustilo RB. Infection after total hip arthroplasty. J Bone Joint Surg [Am] 1996;78:512–23.

[14] Teller RE, Christie MJ, Martin W, et al. Sequential indium-labeled leukocyte and bone scans to diagnose prosthetic infection. Clin Orthop Relat Res 2000;373:241–7.

[15] Magnuson JE, Brown ML, Hauser MR, et al. In-111-labeled leukocyte scintigraphy in suspected orthopedic prosthesis infection: comparison with other imaging modalities. Radiology 1988;168:235–9.

[16] Cuckler JM, Star AM, Alavi A, et al. Diagnosis and management of the infected total joint arthroplasty. Orthop Clin North Am 1991;22:523–30.

[17] Perry CR, Pearson RL, Miller GA. Accuracy of cultures of material from swabbing of the superficial aspect of the wound and needle biopsy in the preoperative assessment of osteomyelitis. J Bone Joint Surg Am 1991;73:745–9.

[18] Capitano M, Kirkpatrick JA. Early roentgen observations in acute osteomyelitis. AJR AM J Roentgenol 1970;108:488–96.

[19] Munk PL, Vellet AD, Levin MF, et al. Imaging after arthroplasty. Can Assoc Radiol J 1994;45: 6–15.

[20] Lyons CW, Berquist TH, Lyons JC, et al. Evaluation of radiographic findings in painful hip arthroplasties. Clin Orthop 1985;195:239–51.

[21] Ledermann HP, Morrison WB, Schweitzer ME. MR image analysis of pedal osteomyelitis: distribution, patterns of spread, and frequency of associated ulceration and septic arthritis. Radiology 2002;223:747–55.

[22] De Winter F, Vogelaers, Gemme F, et al. Promising role of 18-F-fluoro-D-deoxyglucose positron emission tomography in clinical infectious diseases. Eur J Clin Microbiol Infect Dis 2002; 21:247–57.

[23] Nadel HR, Stilwell ME. Nuclear medicine topics in pediatric musculoskeletal disease: techniques and applications. Radiol Clin North Am 2001; 39:619–51.

[24] Palestro CJ, Love C, Tronco GC, et al. Role of radionuclide imaging in the diagnosis of postoperative infection. Radiographics 2000;20: 1649–60.

[25] Peters AM. The use of nuclear medicine in infections. Br J Radiol 1998;71:252–61.

[26] Teller RE, Christie MJ, Martin W, et al. Sequential indium-labeled leukocyte and bone scans to diagnose prosthetic joint infection. Clin Orthop 2000;373:241–7.

[27] Corstens FH, van der Meer JW. Nuclear medicine's role in infection and inflammation. Lancet 1999;354:765–70.

[28] Gratz S, Rennen HJ, Boerman OC, et al. 99mTc-HMPAO-labeled autologous versus heterologous leukocytes for imaging infection. J Nucl Med 2002;43:918–24.

[29] Van Acker F, Nuyts J, Maes A, et al. FDG-PET, 99mtc-HMPAO white blood cell SPET and bone scintigraphy in the evaluation of painful total knee arthroplasties. Eur J Nucl Med 2001;28: 1496–504.

[30] Chacko TK, Zhuang H, Stevenson K, et al. The importance of the location of fluorodeoxyglucose uptake in periprosthetic infection in painful hip prostheses. Nucl Med Commun 2002;23:851–5.

[31] Thakur ML, Marcus CS, Kipper SL, et al. Imaging infection with LeuTech. Nucl Med Commun 2001;22:513–9.

RADIOLOGIC
CLINICS
OF NORTH AMERICA

Radiol Clin N Am 44 (2006) 451–461

Bone Graft Materials and Synthetic Substitutes

Francesca D. Beaman, MD[a], Laura W. Bancroft, MD[a,*],
Jeffrey J. Peterson, MD[a], Mark J. Kransdorf, MD[a,b]

- Principles
- Classification of bone graft materials
  Allografts
  Autografts
  Synthetics
- Summary
- References

Bone graft materials are gaining wide acceptance through their vast availability and usefulness in myriad reconstructive procedures for articular and osseous defects. Functions of bone graft materials include promoting osseous ingrowth and bone healing, providing a structural substrate for these processes, and serving as a vehicle for direct antibiotic delivery. Radiologists are tasked with the need to familiarize themselves with the varied appearances of bone graft materials to avoid misinterpretation of graft material as residual or recurrent disease.

Principles

The primary function of bone graft material is to augment osseous healing by providing a cellular milieu for new bone formation and a structural framework during healing. There are three key concepts in understanding graft function: osteogenesis, osteoinduction, and osteoconduction.

Graft osteogenesis is the transplant of osteogenic precursor cells, capable of new bone formation, that may originate from a graft or host bed. The cellular elements within a graft must be transferred, maintain viability until completion of osseous incorporation, and produce new bone at a recipient site. Types of grafts capable of invoking osteogenesis include cortical and cancellous bone, vascularized bone segments, and bone marrow aspirate [1,2]. Osteoinduction is the method by which pleuripotential mesenchymal cells are recruited from a surrounding host tissue to differentiate into osteoblasts. This transformation is mediated in part by growth factors within a graft, specifically bone morphogenic proteins, which are glycoproteins located in the bone matrix [1–3]. Lastly, a graft may function as a scaffold and facilitate ingrowth of vessels and migration of host cells capable of osteogenesis, a process termed osteoconduction. As new bone is formed, a graft may be resorbed partially or completely (creeping substitution) [2].

Successful incorporation of bone graft material in any site is dependent on new bone formation, structural incorporation of host bone provided by a graft, and adaptive remodeling of the skeleton in

The opinions and assertions contained herein are the private views of the authors and are not to be construed as official or as reflecting the views of the Department of the Army or the Department of Defense.
[a] Department of Radiology, Mayo Clinic, 4500 San Pablo Road, Jacksonville, FL 32224-3899, USA
[b] Department of Radiologic Pathology, Armed Forces Institute of Pathology, Walter Reed Army Medical Center, Building #54, 6825 16th Street, NW, Washington, DC 20306-6000, USA
* Corresponding author.
E-mail address: bancroft.laura@mayo.edu (L.W. Bancroft).

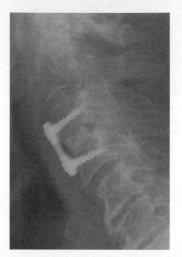

Fig. 1. Unincorporated allograft spacer. A 71-year-old man presents 1 year after anterior cervical discectomy and fusion at C3-C4 using machined allograft spacer and Atlantis plate. Lateral radiograph delineates the persistent distinctiveness of the allograft, indicative of failure of osseous incorporation.

response to mechanical stress [4]. These processes proceed in sequential phases, similar to the principle of fracture healing. The length of time until incorporation is multifactorial and in part reliant on overall patient condition, native bone environment, and properties of the specific graft material [2].

There are many clinical applications for bone graft materials, including (1) osseous fusions—spinal and extremity arthrodesis; (2) fracture stabilization—acute, delayed union, and nonunion; and (3) osseous defects—cavitary, segmental, osteochondral, and arthroplasty related [4]. General complications of all bone graft materials include osseous nonunion, delayed union, graft fracture, graft extrusion, and wound infection. Persistent lucency at a graft-host interface with associated host bone sclerosis, erosion, and fragmentation are radiographic indicators of osseous nonunion.

Classification of bone graft materials

The three primary types of bone graft materials are allografts (graft from cadaveric bone stock), autografts (graft from patients' own bone stock), and synthetic bone graft substitutes [5].

Allografts

Allografts and cadaveric bone transplants provide mainly an osteoconductive or structural matrix and lack osteoinductive properties. Histologically, a initial cellular infiltrate is inflammatory [6]. Subsequently, a graft is surrounded by fibrovascular granulation tissue, which promotes vascular and

osteogenic precursor cell invasion. The interface between allograft transplant and host fibrovascular tissue represents the site of osteoclastic activity and bone resorption. For graft incorporation to occur, the balance between osteoclastic and osteoblastic activity must be maintained [7]. Clinically, examples of allograft usefulness are (1) bone void filler, (2) in vivo antibiotic delivery system, (3) composite allografts, (4) osteoarticular allografts, and (5) onlay grafts [8–11]. Limitations of allografts include the possibility of disease transmission (negligible because of tissue processing and sterilization procedures), procurement cost, host immune response, and inconsistent graft incorporation [Fig. 1] [1,2,12].

Osteoarticular allografts [Fig. 2] are used in abnormalities involving subchondral bone and overlying cartilage, such as avascular necrosis, infection, trauma, osteochondritis dissecans, and neoplasms. They provide an alternative to arthroplasty and to the limited life span of prosthesis in a young patient. Three basic forms are used: osteochondral shell allografts (combination of subchondral bone and adjacent cartilage), half-joint allografts, and whole-joint allografts (used primarily for reconstruction after neoplasm resection). Osteoarticular grafts allow bone healing and revascularization

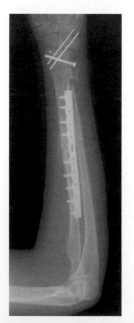

Fig. 2. Osteochondral allograft placed in a 24-year-old woman who had right distal radial grade 1 osteosarcoma. Lateral postoperative radiograph shows placement of the radial osteochondral allograft with mid and distal radial plate and screw fixation, two K-wires through the radiocarpal joint, and suture anchor in the radial styloid. Note the osteotomy site between the native radius and the allograft.

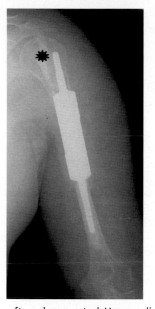

Fig. 3. Allograft and cemented Howmedica humeral intercalary segmental replacement prosthesis. Anteroposterior (AP) radiograph shows the composite allograft (*) and prosthesis used to reconstruct the humerus after en bloc resection of the diaphysis secondary to metastatic adenocarcinoma with extensive bone destruction in a 47-year-old man.

before permanent cartilage and osseous destruction occur. The success of these grafts is reliant mainly on the osseous portion, as it is the weakest region during revascularization. Radiographically, the anatomic alignment of a graft should be assessed with the native articular surface to ensure congruence and graft incorporation. Graft failure may be diagnosed if there is radiographic dissociation or malalignment of a graft from the native articular surface; associated fragmentation, persistent graft opacity, and increased bone resorption also may be present [6].

Onlay or strut allografts function as long bone scaffolding, for example, in cases of periprosthetic fractures or large en bloc resections [13]. Strut graft complications include fracture nonunion and graft fracture. In surgical reconstructions, bone allografts often are placed in tandem with implants and fixation devices. Graft failure, however, allows excessive mechanical force to be placed on the hardware, which consequently may result in hardware failure.

Composite allografts are combinations of graft materials that provide optimal reconstruction and structural support. Allograft-prosthesis composites are used after primary musculoskeletal tumor resection and limb salvage and can provide stable and mobile joints [Fig. 3]. Composite allografts also are used for distal humerus reconstruction, in anterior cruciate ligament grafting with semitendinosus

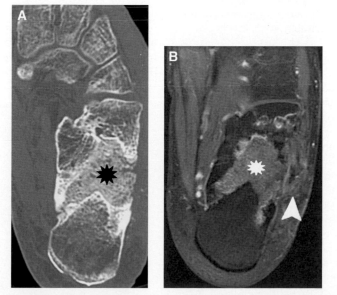

Fig. 4. Enhancement of cancellous allograft may mimic osteomyelitis. (*A*) Axial CT image shows dense graft material (*) in the left calcaneus, which was placed after fracture through the midportion of the calcaneus with extension into the posterior facet of the subtalar joint in a 63-year-old man. CT was performed 15 months postoperatively. (*B*) Axial spin-echo, fat-saturation, T1-weighted MR image with contrast enhancement (repetition time [TR] 516/echo time [TE] 14) shows abnormal signal within the calcaneus with minimal low-level enhancement (*). Enhancement of the adjacent soft tissues (*arrowhead*) is consistent with cellulitis. Given the cellulitis and graft enhancement, superimposed osteomyelitis could not be excluded. Subsequent operative cultures were negative.

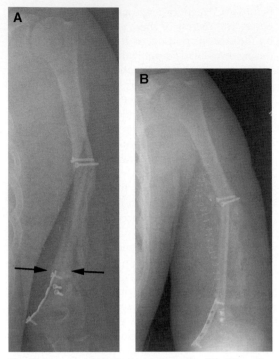

Fig. 5. Vascularized fibular autograft. (*A*) AP radiograph of the left humerus in a 20-year-old man who suffered grade 2 open fracture secondary to a motor vehicle accident and subsequent fracture nonunion and osteomyelitis. Six days after placement of a vascularized fibular autograft (*between arrows*) bridging the remaining humerus. (*B*) One-year postoperative AP radiograph shows complete incorporation of the autograft and graft hypertrophy.

bone composite allograft with mixed corticalcancellous bone dowels at each end, and in a variety of other anatomic locations [14].

Cortical allografts have opacity similar to native cortical bone on radiography and CT. In an initial postoperative period, strut grafts appear as tubular bones with defined cortices and a medullary canal. Chip or morcellized forms of allograft do not retain these characteristics, but appear as an opaque masslike conglomerate within the bony defect. The discrete boundary between host and graft easily is identifiable in the initial period. As union progresses, this junction becomes obliterated secondary to osseous trabecular ingrowth and medullary canal replacement by fibrous tissue.

MR imaging may be useful in evaluating allograft incorporation or failure by delineating graft marrow signal intensity. In the immediate postoperative period, allografts show hypointense signal on all MR pulse sequences. Ingrowth of hematopoietic tissue replaces graft fatty marrow in the later phases of incorporation, manifested by the presence of red marrow signal. Persistent T1- and T2-weighted signal hypointensity implies fibrous replacement of

the medullary canal and lack of complete graft incorporation [6,15]. Mild, homogeneous allograft enhancement may occur and should not be misinterpreted as pathology [Fig. 4].

Autografts

Autogenous cortical, cancellous, or corticocancellous bone and bone marrow aspirate have the main advantages of supplying bone volume and osteogenic cells capable of new bone formation, as they incorporate all three properties of osteogenesis, osteoinduction, and osteoconduction [1,12,16]. Autografts are harvested primarily from the iliac crest and fibula, but they also may be acquired from the other long bones, femoral head, or ribs. Physical forms include paste, morsels, chips, strips, matchsticks, block, and segments [4].

Vascularized bone grafts represent a subset of autografts, which require harvesting the bone of interest and relevant blood supply. The three most common donor sites are ribs (posterior intercostals vessels), the iliac crest (deep circumflex iliac vessels), and fibula (peroneal vessels). Preoperative arteriograms of donor and recipient sites are advocated widely to allow for identification of any anatomic arterial variations. Vascularized bone grafts augment healing by improving the overall survival of osteocytes and enhancing osteogenesis. Radiographically, the margins of a graft fade as trabecular ingrowth progresses. With further osseous union, there is a progressive diminution in the overall opacity of a graft, with an increase in the opacity of recipient site secondary to new bone formation [6]. Vascularized graft hypertrophy can occur [Fig. 5], especially in weight-bearing portions of

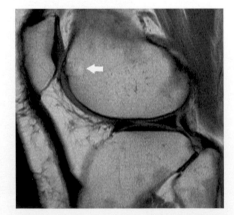

Fig. 6. Osteochondral transfer right lateral femoral condyle. Sagittal turbo spin-echo proton density MR image (TR 3000/TE 26) after osteochondral autograft transfer (*arrow*) in a 47-year-old woman shows the osseous contour of the graft plug extends beyond femoral contour; however, the cartilage is flush with the adjacent normal cartilage.

the body. Technetium 99m three-phase bone scan also is a sensitive technique available for the evaluation of graft viability. In the immediate postoperative period, radionuclide accumulation is consistent with an intact vascular supply and active bone metabolism. After approximately 1 week, however, activity also may be the result of osteoblastic activity [6]. Overall, a vascularized fibular graft should incorporate in approximately 3 to 5 months [6].

Focal chondral and osteochondral defects located on the articular weight-bearing surfaces of bones represent problematic orthopedic lesions, usually causing significant patient morbidity [8]. Histori-

cally, several orthopedic techniques are used in treatment of these lesions, including arthroscopic debridement, abrasion arthroplasty, subchondral drilling/microfracture, periosteal grafts, autogenous chondrocyte implantation, and autologous osteochondral mosaicplasty [Fig. 6] [17–26]. Mosaicplasty is a procedure in which small, cylindric osteochondral grafts are harvested from the non–weight-bearing portions of femoral condyles, typically at the level of the patellofemoral joint, and transplanted into the critical lesion [26]. Over time, graft incorporation occurs with native bone.

Rarely, a segment of tumorous bone may be removed from a patient, autoclaved or irradiated,

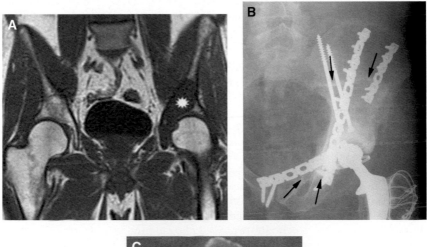

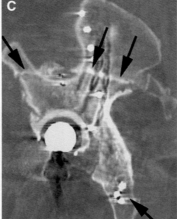

Fig. 7. Autoclaved allograft placed in a 46-year-old man who had left periacetabular region grade 1 chondrosarcoma. (*A*) Coronal spin-echo, T1-weighted MR image (TR 350/TE 10) shows the abnormal hypointense signal (*star*) from the chondrosarcoma in the left acetabulum with extension into the ipsilateral proximal superior ramus and the iliac wing. (*B*) AP radiograph performed 1 year and 11 months after left hemipelvectomy; reconstruction with autoclaved autograft (*arrows*) and total hip arthroplasty shows overall graft density to be comparable to native bone. Allografts were unsatisfactory in this case because they failed to provide the proper acetabular angle, risking femoral malalignment. Increased periacetabular density is cement from the total hip arthroplasty. (*C*) Sagittal reconstruction CT image performed 3 years and 9 months postoperatively shows the density of the graft (*arrows*) comparable to native bone. CT offers an effective means by which to evaluate the graft for new areas of sclerosis or lysis.

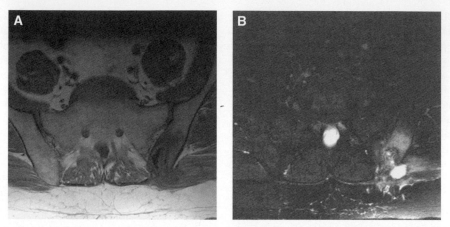

Fig. 8. Infected posterior iliac crest morcellized bone graft harvest site in a 56-year-old man. (*A*) Axial, spin-echo, T1-weighted (TR 500/TE 12) and (*B*) axial, fast spin-echo, fat-saturated, T2-weighted (TR 4000/TE 102) MR images, obtained 3 months postoperatively, show the surgical defect in the posterior left ilium, which has a complex appearance with areas of increased and decreased signal and significant associated edema. After contrast administration there were loculated nonenhancing areas.

and returned to the donor site. This technique may be used, for example, when the specific contours of a cadaveric allograft cannot match a particular patient's innominate bone optimally [Fig. 7].

The imaging appearances of autografts are dependent on type (ie, block or segments of bone versus morcelized chips), composition (cortical, cancellous, corticocanellous bone, or bone marrow aspirate), and age of a graft. For example, chip autografts initially look like osseous fragments on radiographs and CT. In contradistinction, healed osteochondral transfer autografts barely are perceptible on conventional radiographs and show minimal MR signal heterogeneity, being nearly isointense to the adjacent marrow and cartilage. Vascularized fibular autografts maintain an appearance identical to a tubular bone with well-defined cortices and medullary canal.

On radiography and CT, autograft opacity is similar to that of adjacent cortical bone. Autografts have a variable postoperative MR imaging appearance, however. On T1-weighted MR imaging, solid bony fusion is evident, as normal marrow signal bridges graft and host bone with uninterrupted cortices and a single medually canal. Alternatively, viable marrow in an autograft may be hyperintense to skeletal muscle on T1-weighted sequences and hypointense on T2-weighted MR imaging sequences, features not present with other graft materials. Hypointense T1-weighted signal in conjunction with isointense to hyperintense T2-weighted signal correlates histologically with graft necrosis and ingrowth of granulation tissue [16]. Vascularized fibular autografts should maintain T1- and T2-weighted marrow signal intensity similar to that of host bone marrow; without these similarities, vascular and, therefore, graft compromise may be suspected [6,27,28].

In an effort to acquire osteoconductive and osteoinductive properties, combinations of graft materials may be created. Autologous platelet rich plasma (PRP) is a concentration of human platelets within a small volume of plasma. Functionally, PRP is an osteoinductive substance that initiates wound healing by supplying fundamental growth protein factors secreted by platelets and cell adhesion molecules fibrin, fibronectin, and vitronectin [29]. Therefore, PRP mixed with allografts or autografts affords structural integrity and growth factors. Radiographically, only opaque allograft or autograft material is detectable, as PRP is lucent.

A subset of complications specific to autografts includes those related to graft procurement donor sites, specifically, increased operative time, increased blood loss, wound complications [Fig. 8],

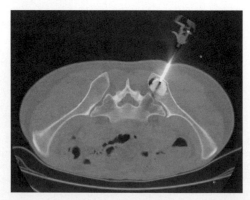

Fig. 9. Minimally Invasive Injectable Graft (MIIG) (Wright Medical Technology, Arlington, Tennessee). Axial CT image during injection of MIIG used to fill the void from a unicameral bone cyst in the right poserior ilium in a 42-year-old man.

chronic pain, scarring, local sensory loss, and limited availability of suitable host donor sites.

Synthetics

Synthetic bone graft substitutes, alternatives to allografts and autografts, may be subdivided into three primary groups: demineralized bone matrix (DBM), ceramics, and composite grafts [2]. In an effort to bypass the drawbacks inherent to allografts and autografts, use of synthetic substitutes has intensified. Synthetic substitutes are available in a variety of formulations, including pellets, cement [Figs. 9 and 10], and injectable paste [Fig. 11].

DBM is created through an acid extraction process from cortical or corticocancellous bone producing a composite of noncollagenous proteins, bone growth factors, and collagen [30]. Demineralized bone implants have osteoinductive properties, stimulating osseous healing in 3 to 6 months after surgery, and reportedly show no significant graft resorption at 7 years [31,32]. Disadvantages of demineralized bone are loss of structural rigidity resulting from processing and inability to localize the material radiographically because of its inherent lucency [33].

One example of demineralized bone is Grafton DBM (Osteotech, Eatonton, New Jersey), which may be packed into bone voids. Graft material only barely is perceptible on postoperative radiographs. Grafton also may serve as a bone graft extender and be implanted as a mixture with cancellous allograft to obtain osteoconductive and osteoinductive properties. The combination of a DBM and allograft chips has a visual opacity between that of cortical and medullary bone on

Fig. 11. Demineralized Bone Matrix (DBX) (Dentsply FC, Lakewood, Colorado) paste. Product photograph after the paste is dispensed from prepackaged syringe. (*Courtesy of* Synthes USA, Paoli, PA; with permission.)

radiographs [Fig. 12] and CT [Fig. 13]. Diffuse enhancement on MR imaging within the postoperative bed is an expected postoperative finding, presumably the result of the composition of graft material in conjunction with the ingrowth of vascularized granulation tissue.

The classification scheme of ceramics is based on graft composition and may be divided into hydroxyapatite, tricalcium phosphate (TCP), biphasic calcium phosphate (combination of hydroxyapatite and TCP), and calcium sulfate. Ceramics provide an osteoconductive lattice on which host osteogenesis may occur; however, they lack osteoinductive properties [2]. Ceramics are formulated

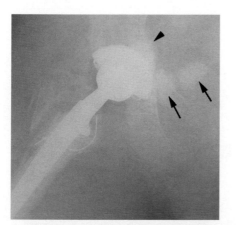

Fig. 10. MIIG extravasation. Postoperative AP radiograph obtained immediately after curettage of osteolytic acetabular defect secondary to particle disease with injection of MIIG filling the osseous defects (*arrowhead*). Note the MIIG extravasation (*arrows*) medial to the acetabulum and iliac bone secondary to the filling of uncontained osseous defects.

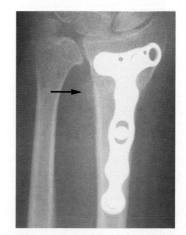

Fig. 12. Open reduction internal fixation left distal radius fracture with composite of DBX putty and cancellous allograft chips. AP radiograph performed 1 month postoperatively shows AcuMed dorsal plate transfixing a distal radius fracture in a 69-year-old woman. Mottled radiodensities (*arrow*) represent cancellous allograft chips and DBX putty mixture.

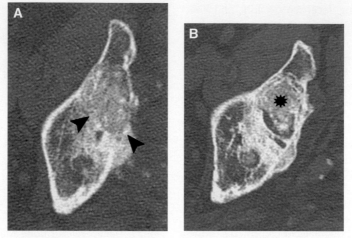

Fig. 13. CT series of cancellous allograft with Grafton gel (Osteotech, Eatonton, New Jersey). (*A*) Axial CT image of the left hip obtained on postoperative day 1 shows bone graft material completely filling the superomedial acetabular void resulting from polyethylene osteolysis in a 76-year-old man who had a total hip arthroplasty (*arrowheads*). (*B*) Axial CT image obtained 1-year post procedure shows incorporation of the graft with persistence of the dense material (*) in the superior acetabulum. Note is made of minimal osteolysis adjacent to zone I of the acetabular component.

for use in spaces that are not intrinsic to host bone stability. The product design allows for creeping substitution, which involves osseous resorption and replacement as part of the healing process.

Periprosthetic infection can be treated with local antibiotic delivery vehicles, such as antibiotic-loaded bone cement [Fig. 14]. This combination of antibiotics and ceramics is the gold standard for local antibiotic delivery, because it can deliver high concentrations of antibiotics (ie, gentamicin and tobramycin) to avascular areas that are inaccessible by systemic antibiotics [34].

In an initial postoperative period, ceramics are more opaque than adjacent native bone on radiography and CT [**Figs. 15 and 16**], with Hounsfield units similar to that of cortical bone. The margins and internal architecture of graft material are de-

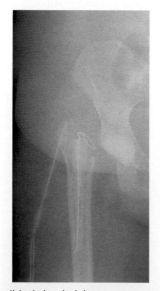

Fig. 14. Antibiotic-loaded bone cement and Girdlestone for infected total hip arthroplasty. A 31-year-old woman acquired *Staphylococcus aureus* infection and required removal of hardware. The antibiotic-loaded (tobramycin, vancomycin, and cefazolin) bone cement in the acetabular fossa, proximal femoral defect and lateral soft tissue sinus tract was able to deliver high concentrations of antibiotics to avascular areas inaccessible by systemic antibiotics.

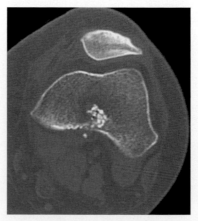

Fig. 15. Conduit TCP pellets (DePuy/Johnson and Johnson, Warsaw, Indiana) mixed with Grafton putty. Axial CT image of the left distal femur postoperative day 1 after enchondroma curettage and insertion of graft material in a 61-year-old woman nicely shows the individual dense pellets filling the bone void.

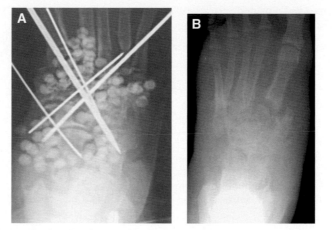

Fig. 16. Osteoset (Wright Medical Technology, Arlington, Tennessee). (*A*) Intraoperative radiograph shows dense Osteoset pellets filling cystic lesions in midfoot and talus resulting from erosions from tophaceous gout in a 24-year-old woman. (*B*) AP radiograph obtained 15 months after surgery shows Osteoset resorption with minimal bony ingrowth.

fined sharply from host bone. As osseous ingrowth proceeds, subsequent radiographic obliteration of a graft-host junction is observed with loss of distinctive implant margins and internal architecture [see Fig. 16]. Pathologically, this evolution is secondary to osteoclastic activity that promotes osseous ingrowth and osseous fusion [6]. Potential pitfalls exist when imaging ceramics by MR, as they may appear hypointense or isointense and mass like on all pulse sequences, features that can be mistaken for residual or recurrent tumor, such as giant cell tumor or pigmented villonodular synovitis [Fig. 17].

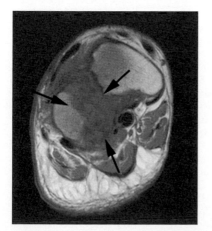

Fig. 17. Osteoset mimics recurrent pigmented villonodular synovitis (PVNS) in a 51-year-old woman. Axial, spin-echo, T1-weighted MR image (TR 400/TE 17) shows large ill-defined mass-like area (*arrows*) in the midfoot surgical bed and demonstrates intermediate to low signal on all pulse sequences, mimicking recurrent PVNS.

Composite grafts incorporate the advantages of DBM and ceramics into a single compound. DBM affords osteoinductive properties, whereas ceramics provide an osteoconductive structural matrix. Analogous mixtures also may be created at the surgical table to accomplish similar goals. As is the case with other synthetics, composites are opaque on radiography and CT and show gradual incorporation.

Summary

Allografts, autografts, and synthetic substitutes are widespread in orthopedic surgery, and radiologists must be familiar with bone graft categories, functions, and varied imaging appearances. Bone allografts and autografts initially are opaque on radiography and CT. On sequential examinations, these materials may remain opaque or show a spectrum to lucency, as creeping substitution allows for gradual graft resorption with osseous ingrowth. In the immediate postoperative period, allografts show hypointensity on T1- and T2-weighted pulse sequences. Graft incorporation or failure may be diagnosed by the dynamic or static nature of allograft marrow signal intensity, respectively. In contradistinction, autografts have a variable postoperative MR appearance. Radiographically, ceramics are more opaque than the adjacent native bone. Care must be taken in the interpretation of postoperative MR images, as mild enhancement of graft material may be seen in normal conditions. Complications of graft placement include early graft resorption, osseous nonunion, wound infection, and infectious disease transmission. As synthetic substitutes emerge and evolve, recognition of their expected imaging characteristics will become critical in an effort to avoid diagnostic pitfalls.

References

[1] Greenwald AS, Boden SD, Goldberg VM, et al. Bone-graft substitutes: facts, fictions, and applications. J Bone Joint Surg [Am] 2001;83-A: 98–103.

[2] Lind M, Bunger C. Factors stimulating bone formation. In: Gunzburg R, Szpalski M, Passuti N, et al, editors. The use of bone substitutes in spine surgery. Berlin (Germany): Springer-Verlag; 2002. p. 18–25.

[3] Urist MR. Bone: formation by autoinduction. Science 1965;150:893–9.

[4] Muschler GF, Lane JM. Orthopedic surgery. In: Habal MB, Reddi AH, editors. Bone grafts and bone substitutes. Philadelphia: WB Saunders; 1992. p. 375–407.

[5] Habal MB. Different forms of bone grafts. In: Habal MB, Reddi AH, editors. Bone grafts and bone substitutes. Philadelphia: WB Saunders; 1992. p. 6–8.

[6] Murphey MD, Sartoris DJ, Bramble JM. Radiographic assessment of bone grafts. In: Habal MB, Reddi AH, editors. Bone grafts and bone substitutes. Philadelphia: WB Saunders; 1992. p. 9–36.

[7] Moran CG, Wood MB. Bone allografts. In: Wood MB, Gilbert A, editors. Microvascular bone reconstruction. St. Louis (MO): Mosby; 1997. p. 41–3.

[8] Gitelis S, Brebach GT. The treatment of chronic osteomyelitis with a biodegradable antibiotic-impregnated implant. Am J Orthop Surg 2002; 10:53–60.

[9] Witso E, Loseth K, Bergh K. Adsorption and release of antibiotics from morselized cancellous bone. In vitro studies of 8 antibiotics. Acta Orthop Scand 1999;70:298–304.

[10] Czitron AA. Allograft reconstruction after tumor surgery in the appendicular skeleton. In: Czitrom AA, Gross AE, editors. Allografts in orthopaedic practice. Baltimore: Williams and Wilkins; 1992. p. 83–119.

[11] Armstrong JR. Types, sources, and fixation of grafts. In: Bone-grafting in the treatment of fractures. Baltimore: Williams and Wilkins; 1945. p. 11–21.

[12] Delloye C, Cnockaert N, Cornu O. Bone substitutes in 2003: an overview. Acta Orthop Belg 2003;69:1–8.

[13] Gross AE. Revision arthroplasty of the hip using allograft bone. In: Czitrom AA, Gross AE, editors. Allografts in orthopaedic practice. Baltimore: Williams and Wilkins; 1992. p. 147–73.

[14] Kocabey Y, Klein S, Nyland J, et al. Tibial fixation comparison of semitendinosus-bone composite allografts fixed with bioabsorbable screws and bone-patella tendon-bone grafts fixed with ti-

tanium screws. Knee Surg Sports Traumatol Arthosc 2004;12:88–93.

[15] Jelinek JS, Kransdorf MJ, Moser RP, et al. MR imaging findings in patients with bone-chip allografts. AJR Am J Roentgenol 1990;155: 1257–60.

[16] Moran CG, Wood MB. Bone autograft incorporation. In: Wood MB, Gilbert A, editors. Microvascular bone reconstruction. St. Louis: Mosby; 1997. p. 33–6.

[17] Buckwalter JA, Mankin HJ. Articular cartilage: degeneration and osteoarthrosis, repair, regeneration, and transplantation. Instr Course Lect 1998;47:487–504.

[18] Hubbard MJ. Articular debridement versus washout for degeneration of the medial femoral condyle. A five year study. J Bone Joint Surg [Br] 1996;78-B:217–9.

[19] Bentley G, Dowd G. Current concepts of the etiology and treatment of chondromalacia patelae. Clin Orthop 1984;189:209–28.

[20] Steadman JR, Briggs KK, Rodrigo JJ, et al. Outcomes of microfracture for traumatic chondral defects of the knee: average 11-year follow-up. Arthroscopy 2003;19:477–84.

[21] O'Driscoll SW, Keeley FW, Salter RB. Durability of regenerated articular cartilage produced by free autogenous periosteal grafts in major full-thickness defects in joint surfaces under the influence of continuous passive motion: a follow up report at one year. J Bone Joint Surg [Am] 1988;70-A:595–606.

[22] Brittberg M, Lindahl A, Nilsson A, et al. Treatment of deep cartilage defects in the knee with autologous chondrocyte transplantation. N Engl J Med 1994;331:889–95.

[23] Hunziker RB, Rosenberg LC. Repair of partial-thickness defects in articular cartilage: Cell recruitment from the synovial membrane. J Bone Joint Surg [Am] 1996;78-A:721–33.

[24] Bentley G, Biant LC, Carrington RWJ, et al. A prospective, randomised comparison of autologous chondrocyte implantation versus mosaicplasty for osteochondral defects in the knee. J Bone Joint Surg [Br] 2003;85-B:223–30.

[25] Evans PJ, Miniaci A, Hurtig MB. Manual punch versus power harvesting of osteochondral grafts. Arthroscopy 2004;20:306–10.

[26] Hangody L, Rathonyi GK, Duska Z, et al. Autologous osteochondral mosaicplasty. Surgical technique. J Bone Joint Surg [Am] 2004;86-A:65–72.

[27] Shaffer JW, Field GA, Goldberg VM. Biology of vascularized bone grafts. In: Habal MB, Reddi AH, editors. Bone grafts and bone substitutes. Philadelphia: WB Saunders; 1992. p. 37–52.

[28] Moran CG, Wood MB. Vascularized autograft healing. In: Wood MB, Gilbert A, editors. Microvascular bone reconstruction. St. Louis: Mosby; 1997. p. 37–9.

[29] Marx RE. Platelet-rich plasma: evidence to support its use. J Oral Maxillofac Surg 2004;62:489–96.

[30] Ross JS, Masaryk TJ, Modic MT, et al. Lumbar

spine: postoperative assessment with surface-coil MR imaging. Radiology 1987;164:851–60.

[31] Martin G, Boden S, Titus L, et al. New formulations of demineralized bone matrix as a more effective graft alternative in experimental posterolateral lumbar spine arthrodesis. Spine 1999;24: 637–45.

[32] Lindholm TS, Ragni P. Experimental spinal fusion using demineralized bone matrix, bone marrow and hydroxyapatite. In: Aebi M, Regaz-zoni P, editors. Bone transplantation. Berlin (Germany): Springer-Verlag; 1989. p. 227–8.

[33] Glowacki J. Cellular responses to bone-derived materials. In: Friedlaender GE, Goldberg VM, editors. Bone and cartilage allografts: biology and clinical applications. Warrenton (VA): Airlie House; 1991. p. 55–73.

[34] Wang J, Calhoun JH, Mader JT. The application of bioimplants in the management of chronic osteomyelitis. Orthopedics 2002;25:1247–52.

ELSEVIER
SAUNDERS

RADIOLOGIC
CLINICS
OF NORTH AMERICA

Radiol Clin N Am 44 (2006) 463–472

Soft Tissue Tumors: Post-Treatment Imaging

Mark J. Kransdorf, MD[a,b,c,*], Mark D. Murphey, MD[c,d,e]

- Clinical history
- Radiographic evaluation
- Imaging evaluation
- Local recurrence
- Post-treatment changes
  Chemotherapy
  Radiation therapy

- *Postoperative fluid and hemorrhage*
- Reconstructive surgery
- Summary
- References

Approximately half of patients who have soft tissue sarcomas have local recurrence [1–3]. Consequently, routine follow-up is essential. As with patients presenting for the initial evaluation of a mass, a systematic approach to the imaging of patients after treatment is essential. It begins with a thorough clinical history, continues with a review of the radiographs, and concludes with a review of the MR imaging. Examination of the operative site also may provide considerable additional information.

Clinical history

The clinical history is critically important in evaluating patients after treatment and, as in the evaluation of an initial lesion, may be the key factor in

establishing an accurate diagnosis. In many circumstances, an accurate history may provide information that allows a specific diagnosis when imaging is nonspecific. Important information includes identification of the original tumor diagnosis. Certain tumors show greater predilection for local recurrence. Fibromatosis, for example, is reported to recur locally in up to three quarters of patients [4,5]. What was the tumor grade? High-grade tumors are at much greater risk for local recurrence. Just as grade and type influence recurrence risk, so does location. Deep tumors, and those in which wide margins cannot be obtained, are at greater risk for local recurrence than are superficial lesions. A well-differentiated liposarcoma of the extremity may be "cured" by wide local excision, whereas

The opinions or assertions contained herein are the private views of the authors and are not to be construed as official or as reflecting the views of the Department of the Army or the Department of Defense.

a Mayo Clinic College of Medicine, 200 First Street, SW, Rochester, MN 55905, USA
b Department of Radiology, Mayo Clinic, 4500 San Pablo Road, Jacksonville, FL 32224-3899, USA
c Department of Radiologic Pathology, Armed Forces Institute of Pathology, 6825 16th Street, NW, Building 54, Room 133A, Washington, DC 20306-6000, USA
d Department of Radiology and Nuclear Medicine, Uniformed Services University of the Health Sciences, 4301 Jones Bridge Road, Bethesda, MD 20814-4799, USA
e Department of Radiology, Walter Reed Army Medical Center, 6900 Georgia Avenue, NW, Building 2, Washington, DC 20307-5001, USA
* Corresponding author. Department of Radiology, Mayo Clinic, 4500 San Pablo Road, Jacksonville, FL 32224-4301.
E-mail address: kransdorf.mark@mayo.edu (M.J. Kransdorf).

the same tumor in the retroperitoneum has virtually 100% local recurrence rate [6].

Appropriate clinical history should include the details of surgery. Which type of surgical resection was done? The type of surgical resection has significant implications for the risk of local recurrence, ranging from exceedingly low risk with radical resection (amputation) to exceedingly high with marginal excision. Surgical data should include the details of the surgical margins. Patients who have positive surgical margins may show no radiologic mass or discrete lesion on initial postoperative imaging while having histologic evidence of residual tumor.

It also is important to determine if there has been radiation therapy or reconstructive surgery and, if so, the time course of each in relation to the current study. Radiation and reconstructive surgery influence the postoperative imaging appearance and both demonstrate time-dependent changes [7–9].

Other important information to glean includes any change in the areas of previous surgery. Any new lumps or bumps? A new mass in the face of anticoagulation may suggest the possibility of a hematoma; however, in the authors' experience, such lesions always are suspicious for hemorrhagic tumor. Finally, is there any evidence or history of metastatic disease? Documented metastases indicate a more aggressive biologic potential and serve to raise the index of suspicion for local recurrence.

Examination of the operative site may provide considerable additional information. It allows confirmation of a clinically suspected palpable abnormality. It also allows inspection of the operative bed to identify the scope of the surgical change, the presence of reconstructive procedures, and any associated conditions, such as radiation dermatitis, cellulitis, or cutaneous ulcers [Fig. 1].

Radiographic evaluation

Similar to the evaluation of a primary mass, the radiologic evaluation of postoperative patients begins with a radiograph. Radiographs can show postoperative skeletal deformity or heterotopic ossification, which may be difficult to appreciate on MR imaging and may masquerade clinically as recurrent tumor. This is helpful especially in cases in which the normal marrow signal is altered by posttreatment changes. Radiographs also can reveal soft tissue calcifications that may not be appreciated on MR imaging. CT is a useful adjunct in specific circumstances. The authors generally reserve CT for those patients in whom radiographs do not depict the lesion, its pattern of mineralization, or its relationship to the host osseous structures adequately. This typically is in areas in which the osseous anatomy is complex, such as the pelvis, shoulder, and paraspinal regions [1].

Imaging evaluation

In postoperative patients, the initial step in the imaging evaluation should include a review of the previous MR imaging studies, to include the presenting and pretreatment studies when available. Although not all change is indicative of tumor recurrence, no change is reassuring in excluding recurrence. Additionally, it is the authors' experience that recurrent tumor frequently exhibits an imaging appearance similar to that of the primary tumor—information that also can be useful in distinguishing tumor recurrence from postoperative change.

Although the technical considerations for MR imaging of patients after treatment mirror those for initial tumor evaluation, there are additional considerations. It is essential to evaluate an entire operative or treated area. Markers noting the margins of the surgical scar can be helpful in ensuring complete coverage. Contrast also may be a useful adjunct. Although in general, the authors find that in most cases recurrent tumor usually is well seen without contrast, there are specific instances in which contrast is especially useful; these include the distinction of postoperative hematoma from recurrent tumor and the evaluation of recurrent fibromatosis. In the evaluation of lesions with increased signal intensity on T1-weighted images, comparison of precontrast, fat-suppressed, T1-weighted images with postcontrast, fat-suppressed, T1-weighted images is preferred. Hemorrhage shows increased signal on enhanced fat-suppressed im-

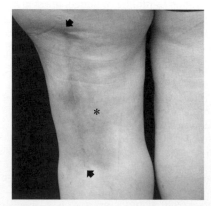

Fig. 1. Clinical examination: clinical photograph of postoperative site. A 60-year-old woman 6-years after resection and radiation therapy for extraskeletal osteosarcoma. Photograph shows scar (*arrows*) and hyperpigmentation (*asterisk*) from radiation.

aging and may be mistaken for enhancement without careful comparison to precontrast imaging. If precontrast, fat-suppressed imaging is not done, comparison with non–fat-suppressed enhanced imaging is useful.

Local recurrence

MR imaging and sonography are shown to be useful in detecting local recurrence [3]. Recurrent tumor is characterized by the presence of a discrete nodule or mass, typically with prolonged T1 and T2 relaxation times, usually demonstrating prominent contrast enhancement [Fig. 2]. Tumor recurrence, however, usually is seen well without enhanced imaging. When identification of a discrete nodule with increased signal density on fluid-sensitive images is used as a criterion for local recurrence, MR imaging is quite accurate [10].

Postsurgical changes are more variable and usually show areas of low or intermediate signal intensity or fluid collections, without a discrete nodule [3,10]. The spectrum of post-treatment changes is described later more fully. When a mass with high signal intensity on T2-weighted images is found, tumor must be differentiated from postoperative hygroma. In most cases, this differentiation is straightforward with fluid characterized by a homogeneous, well-defined mass with prolonged T1- and T2-weighted spin-echo MR images.

As discussed previously, comparison with pretreatment images is essential. For example, myxoid tumors may mimic cysts on MR imaging; hence, differentiation between postoperative fluid collection and recurrent tumor is considerably more difficult if a patient's original tumor was a myxoid liposarcoma. When there is question as to whether or not an area of high signal intensity on T2-

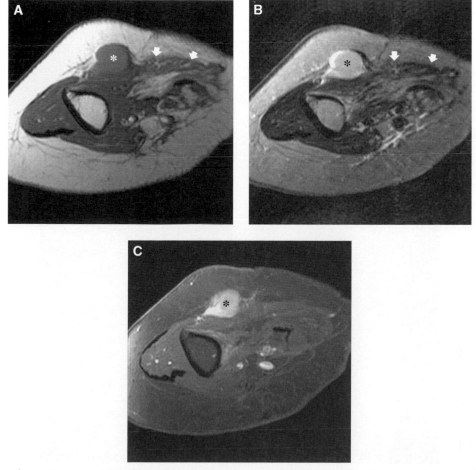

Fig. 2. Local recurrence: recurrent high-grade leiomyosarcoma of the forearm in a 61-year-old woman. (*A*) Axial T1-weighted (repetition time [TR]/echo time [TE]; 535/17) and (*B*) T2-weighted (TR/TE; 2900/83) spin-echo MR images show focal mass (*asterisk*) arising in the operative bed. Note myocutaneous flap (*arrows*). (*C*) Axial fat-suppressed T1-weighted (TR/TE; 539/17) spin-echo MR image after intravenous contrast shows intense enhancement (*asterisk*).

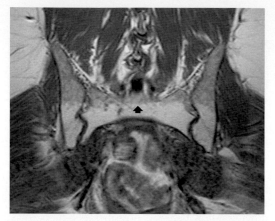

Fig. 3. Radiation therapy: osseous radiation change in the sacrum of a 51-year-old man after radiation for colorectal carcinoma. Coronal T1-weighted (TR/TE; 406/17) spin-echo MR image of the pelvis shows increased fatty marrow in distal sacrum with radiation port well demarcated (*arrow*).

weighted images represents fluid, sonographic examination is an ideal method for further evaluation. It is easy, cheap, and highly accurate. Alternatively, gadolinium-enhanced MR imaging may be used.

Sonography may be used as the primary modality to follow patients for recurrence, with a discrete hypoechoic mass considered to be recurrent tumor [3]. Sonography still is very operator dependent and of limited value in cases where the osseous anatomy is complex, such as in the shoulder girdle or pelvis. Consequently, the authors prefer MR imaging for identification of recurrent tumor.

Post-treatment changes

There are a number of imaging features that are commonly seen following treatment of a soft tissue tumor. Knowledge of these imaging features minimizes the likelihood that they are misconstrued for recurrent tumor. Post-treatment changes include the effects of chemotherapy, pre- or postoperative radiation on bone and soft tissue, postsurgical fluid collections, hemorrhage, and reconstructive myocutaneous flaps.

Chemotherapy

The effects of chemotherapy often are believed inconsequential; however, significant increase in tumor size secondary to chemotherapy-induced hemorrhage is reported [11]. In other cases, chemotherapy may reduce tumor size significantly, with only degenerated and reactive tissue identified at subsequent surgery [12]. The authors find contrast-enhanced imaging useful in determining the degree of intralesional necrosis, a feature that may be useful in establishing the effectiveness of chemotherapy and tumor biologic potential.

Radiation therapy

It long has been recognized that resection of high-grade sarcomas without adjuvant therapy results in unacceptably high rates of local recurrence [13]. In the 1970s, this led oncologists to evaluate the use of adjuvant radiation therapy and chemotherapy as means to improve functional outcome in limb-sparing surgery [13]. In current practice, radiation

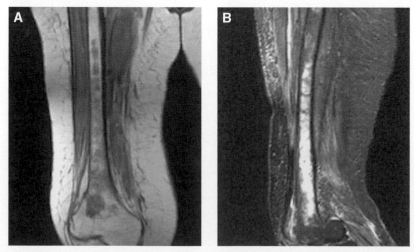

Fig. 4. Radiation therapy marrow changes: high-grade undifferentiated pleomorphic sarcoma (malignant fibrous histiocytoma) in the thigh of an 89-year-old woman treated with 5040-cGy preoperative radiation. (*A*) Coronal T1-weighted and (*B*) axial T2-weighted spin-echo MR images obtained 41 months after radiation therapy show nonspecific abnormal signal within the femoral marrow. Note soft tissue postoperative and postradiation changes.

may be administered before, during, or after tumor surgical resection [14].

Administering radiation after tumor resection avoids the increased risk of wound complications that are associated with operating on compromised tissue. It also eliminates the potential difficulty of determining the histologic diagnosis from radiated tissue. Finally, it allows surgery to be performed without delay [13,14]. Although preoperative radiation is associated with increased wound complication, it reduces tumor volume and, therefore, improves tumor resectability [13,14].

Radiation changes readily are identified in bone and soft tissue. Osseous changes after radiation are well documented; however, this discussion focuses on those changes primarily identified on MR imaging. MR imaging can detect radiation-induced

marrow changes as early as 8 days after onset of therapy [15]. Within 3 and 6 weeks, the marrow shows increasingly heterogeneous signal intensity, with increasing fat signal. In the majority of patients, complete fatty replacement occurs usually within 6 to 8 weeks [**Fig. 3**] [15,16]. Uncommonly, a band of peripheral intermediate signal surrounding central fat may be seen [16]. Approximately half of those patients receiving radiation show marrow changes in the marrow immediately adjacent to the radiated field. These changes are milder and more common in patients who have highly cellular pretreatment marrow [16,17].

As a result of the cellular damage caused by radiation, irradiated bone is at increased risk for fracture [18]. The pelvis is the most likely area to be involved and the incidence of associated pelvic

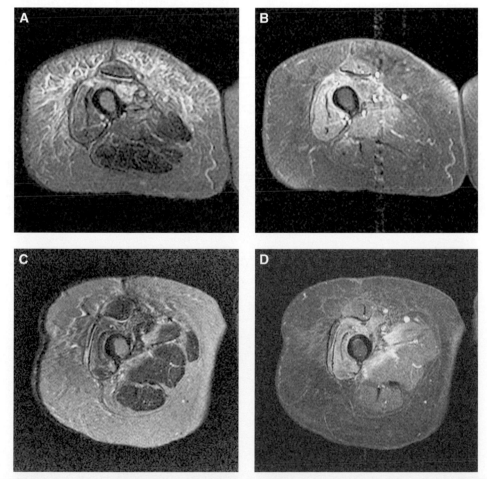

Fig. 5. Radiation therapy soft tissue changes: same patient as in **Fig. 4**. (*A*) Axial T2-weighted (TR/TE; 2800/80) spin-echo MR image obtained 6.5 months following radiation therapy shows extensive edema in the muscle and subcutaneous soft tissues. (*B*) Corresponding axial fat-suppressed T1-weighted (TR/TE; 450/10) spin-echo MR image following contrast shows enhancement of the muscle but little enhancement of the subcutaneous edema. (*C*) Axial T2-weighted (TR/TE; 3262/80) spin-echo MR image obtained 16 months following radiation therapy shows interval improvement. (*D*) Corresponding axial fat-suppressed T1-weighted (TR/TE; 644/15) spin-echo MR image following contrast shows persistent abnormal muscle enhancement.

injury is small—0.1% to 0.3% [19]. This is considerably less than the 1.8% estimated incidence of sacral insufficiency fractures in women ages 55 years and older [20].

The authors also occasionally note the appearance of somewhat poorly defined areas of nonspecific nonadipose tissue within the marrow of radiated extremities. This usually occurs several months after the completion of radiation; moreover, these changes may become more prominent over the succeeding months before stabilizing [Fig. 4].

Soft tissue changes are more variable and differ with type of radiotherapy. Richardson and colleagues [21] characterize the soft tissue changes after radiation and note that irradiated soft tissue shows abnormal edema-like signal intensity. The signal intensity shows great variability but generally was greater on short tau inversion recovery (STIR) images than on T2-weighted spin-echo images. Signal alterations increase with time, being greatest at 12 to 18 months and returning to normal in approximately half of these in 2 to 3 years. The edema signal in the subcutaneous tissue appears as a trabecular or lattice-like pattern of low to intermediate signal intensity on T1-weighted spin-echo images and high signal intensity on fluid-sensitive sequences [11]. Radiation-induced changes in muscle are more diffuse with minimal enhancement after contrast administration and preservation on muscle shape and texture [10,11,22].

MR imaging signal abnormalities are greater and persist longer in the intermuscular septae than in fat or muscle. The size of the fat and intermuscular septae increases mildly, whereas muscle shows a decrease in size after treatment [Fig. 5] [21].

After radiation therapy, patients also may develop inflammatory pseudotumors [Fig. 6]. There is scant documentation of this phenomenon in the imaging literature, but Vanel and coworkers [10] note two such cases in a review of follow-up imaging of 182 patients who had aggressive soft tissue tumors. Both lesions presented as high-signal intensity masses, one 1 year after and the other 12 years after radiation and surgery. Both of these lesions were investigated with dynamic contrast enhancement and showed delayed enhancement (4 to 7 minutes) compared with tumor recurrence (1 to 3 minutes).

Postoperative fluid and hemorrhage

Postoperative fluid collections and hemorrhage after sarcoma surgery show a similar appearance to that seen after nononcology procedures [Fig. 7]. In oncology patients, however, the mass-like appearance of these complications may mimic local tumor recurrence, and the distinction may be difficult especially in selected cases, as in those recurrent tumors complicated by hemorrhage. In these instances, contrast imaging is helpful in making this distinction [Fig. 8].

Uncomplicated postoperative fluid collections are distinguished more readily from tumor recurrence. Recurrent tumor usually is more heterogeneous and the margins usually more irregular than

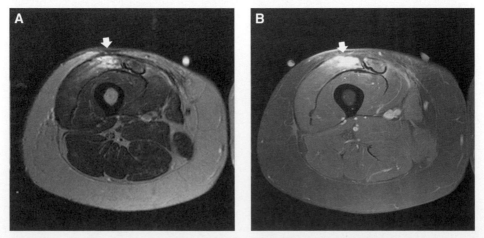

Fig. 6. Radiation pseudotumor: a 61-year-old woman who has a history of a dedifferentiated liposarcoma treated with surgery, chemotherapy, and 4500-cGy radiation therapy. (*A*) Axial T2-weighted spin-echo MR images obtained 3 years after surgery and radiation show a somewhat ill defined area of abnormal signal in the surgical bed (*arrow*), but no discrete mass is identified. (*B*) Axial fat-suppressed T1-weighted (TR/TE; 450/10) spin-echo MR image after contrast shows intense enhancement (*arrow*). Surgical biopsy showed atrophic skeletal muscle and fibrosis.

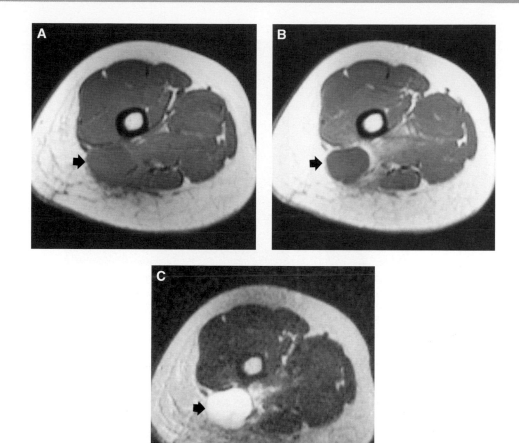

Fig. 7. Postoperative hygroma: MR imaging examination of a 15-year-old girl 3 months after resection of a clear cell sarcoma. Axial T1-weighted (TR/TE; 720/15) spin-echo MR images (*A*) preceding and (*B*) following intravenous contrast administration show a nonenhancing mass-like area (*arrow*) in the posterior thigh. Lesion has a cyst-like margin and shows peripherally ill-defined enhancement. (*C*) Axial T2-weighted (TR/TE; 2180/80) spin-echo MR image shows the lesion to have a "fluid" signal intensity (*arrow*).

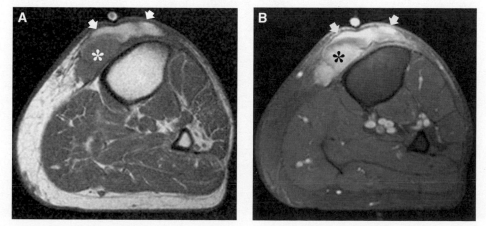

Fig. 8. Postoperative hematoma and tumor recurrence: recurrent leiomyosarcoma in a 66-year-old man after surgical excision complicated by a large hematoma. (*A*) Axial T1-weighted (TR/TE; 740/15) spin-echo image shows a small residual hematoma (*arrows*) with an adjacent mass (*asterisk*). (*B*) Axial fat-suppressed T1 (TR/TE; 475/15) spin-echo MR image after intravenous contrast shows marked enhancement to the recurrent tumor (*asterisk*). Note high signal intensity of subacute blood on fat-suppressed image (*arrows*).

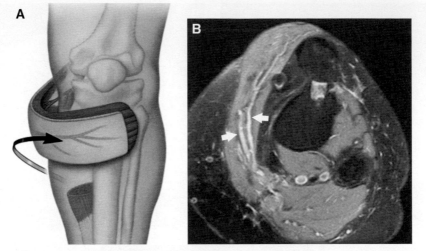

Fig. 9. Rotational flap: 41-year-old woman who has rotational medial gastrocnemius flap used for coverage after debridement for osteomyelitis. (*A*) Diagram of rotational gastrocnemius flap. (*B*) Axial proton-density (TR/TE; 4000/15) image obtained 35 months after surgery demonstrates preserved vascular pedicle (*arrows*) and flap signal characteristics similar to that of background muscles.

those in simple postoperative hygroma. Contrast-enhanced imaging may prove helpful in cases in which seromas demonstrate increased (intermediate) signal intensity on T1-weighted images [10]. Most seromas resolve in 3 to 18 months [12], although this is variable, and they may persist for an extended period of time.

Reconstructive surgery

The extensive tissue resection required to achieve adequate surgical margins in oncologic surgery often requires soft tissue reconstructive surgery. Myocutaneous flaps are used in more than two thirds of extremity sarcoma surgeries [23].

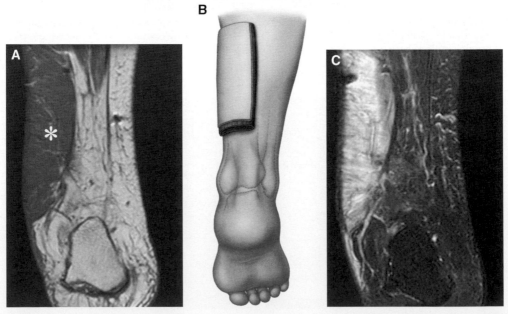

Fig. 10. Free flap: a 39-year-old woman who has a rectus abdominus free flap to ankle after resection of dermatofibrosarcoma protuberans. (*A*) Coronal T1-weighted (TR 683/TE 17) spin-echo MR image at 4 months after placement of free flap (*asterisk*). (*B*) Diagram of free flap. (*C*) Coronal fast spin echo (FSE) T2-weighted (TR/TE; 3000/95) MR image demonstrates markedly increased T2-weighted signal at 4 months after flap placement. Signal intensity returns to that of adjacent muscle in approximately one third of patients.

Myocutaneous flaps contain muscle and overlying skin. Rotational flaps are rotated into position, covering the soft tissue defect while preserving the native neurovascular supply via a pedicle [Fig. 9]. Free flaps are detached completely, placed into the soft tissue defect, and the vascular pedicle is reanastomosed using a microvascular technique [Fig. 10]. Although most flaps are used for coverage only, rotational flaps using muscles, such as the latissimus dorsi, may provide coverage and function when they are used for the upper arm.

As the MR imaging appearance of radiated tissue changes with time, so does the appearance of myocutaneous flaps. A recent report by Fox and colleagues [24] reviews the MR imaging findings in 30 myocutaneous flaps. They note that all flaps demonstrate time-dependent changes in size, signal intensity, and enhancement. All flaps atrophy with time, demonstrating decreased muscle mass and progressive fatty replacement. The amount of atrophy is variable, however, ranging from mild to marked, but is less in flaps providing coverage and function [Fig. 11]. In addition, all flaps initially demonstrate increased signal intensity on T2-weighted images, which return to baseline (similar to the signal intensity of the surrounding muscle) in one third of cases in between 5 and 21 months. Finally, enhancement is seen in approximately three

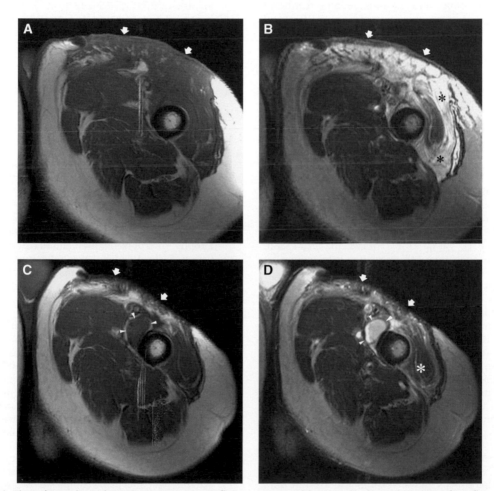

Fig. 11. Time-dependent changes myocutaneous flap: a 42-year-old man who has a latissimus free flap to the upper thigh after surgery and radiation therapy for a malignant fibrous histiocytoma (undifferentiated pleomorphic sarcoma). (*A*) Axial T1-weighted (TR/TE; 410/17) and (*B*) conventional T2-weighted (TR/TE; 2250/80) spin-echo MR images obtained 4 months after surgery. Note the high signal intensity within the flap in (*B*) (*arrows* mark anterior margin of flap) and the high signal intensity in the lateral aspect of the thigh (*asterisks*) from previous radiation. Note muscle texture in flap with curvilinear areas of fat signal in (*A*). (*C*) Axial T1-weighted (TR/TE; 410/17) and (*D*) conventional T2-weighted (TR/TE; 2930/80) spin-echo MR images obtained 31 months after surgery. Note the marked atrophy in the flap (*arrows*). The high signal intensity in the flap and the lateral aspect of the thigh (*asterisk*) has decreased significantly. New recurrence (*arrowheads*) is distinguished easily from post-treatment changes.

quarters of cases, returning to baseline in approximately one third, in 18 months. Postoperative radiation therapy increases the likelihood that a flap exhibits increased signal on T2-weighted images and enhancement.

Summary

In summary, MR imaging is the preferred modality for the evaluation of a soft tissue mass after treatment. Radiographs are useful in identifying skeletal deformity or heterotopic ossification that may masquerade as recurrent tumor. In general, the imaging appearance of recurrent tumor mirrors that of the initial lesion, and the most reliable feature in identifying local recurrence is detection of a focal mass in the tumor bed. Other post-treatment features include post-radiation changes, postoperative fluid and hemorrhage, and postoperative myocutaneous flaps. Careful attention to detail usually allows distinction of recurrent tumor from other post-treatment changes.

References

[1] Kransdorf MJ, Murphey MD. Imaging of soft tissue tumors. In: Imaging of soft tissue tumors. Philadelphia: W.B. Saunders; 1997. p. 37–56.

[2] Vezeridis MP, Moore R, Karakousis CP. Metastatic patterns in soft-tissue sarcomas. Arch Surg 1983;118:915–8.

[3] Choi H, Varma DGK, Fornage BD, et al. Soft-tissue sarcoma: MR imaging vs sonography for the detection of local recurrence after surgery. AJR Am J Roentgenol 1991;157:353–8.

[4] Rock MG, Pritchard DJ, Reiman HM, et al. Extra-abdominal desmoid tumors. J Bone Joint Surg [Am] 1984;66-A:1369–74.

[5] Griffiths HJ, Robinson K, Bonfiglo TA. Agressive fibromatosis. Skeletal Radiol 1983;9:179–82.

[6] Peterson JJ, Kransdorf MJ, Bancroft LW, et al. Malignant fatty tumors: classification, clinical course, imaging appearance and treatment. Skeletal Radiol 2003;32:493–503.

[7] Som P, Urken M, Biller H, et al. Imaging of the postoperative neck. Radiology 1993;187:593–603.

[8] Wester D, Whiteman M, Singer S, et al. Imaging of the postoperative neck with emphasis on surgical flaps and their complications. AJR Am J Roentgenol 1995;164:989–93.

[9] Naidich M, Weissman J. Reconstructive myofascial skull-base flaps: Normal appearance on CT and MR imaging studies. AJR Am J Roentgenol 1996;167:611–4.

[10] Vanel D, Shapeero LG, De Baere T, et al. MR imaging in the follow-up of malignant and aggressive soft-tissue tumors: results of 511 examinations. Radiology 1994;190:263–8.

[11] Varma DGK, Jackson EF, Pollock RE, et al. Soft–tissue sarcoma of the extremities. MR appearance of post-treatment changes and local recurrence. MRI Clin North Am 1995;3:695–710.

[12] Pezzi CM, Pollock RE, Evans HL, et al. Preoperative chemotherapy for soft-tissue sarcomas of the extremities. Ann Surg 1990;476:476–81.

[13] Malawer M, Suarbaker PH. The role of radiation therapy in the treatment of bone and soft-tissue sarcomas. In: Musculoskeletal cancer surgery. Treatment of sarcomas and allied diseases. Dordrecht, The Netherlands: Kluwer Academic Publishers; 2001. p. 85–133.

[14] Khatri VP, Goodnight Jr JE. Extremity soft tissue sarcoma: controversial management issues. Surg Oncol 2005;14:1–9.

[15] Blomlie V, Rofstad EK, Skjonsberg A, et al. Female pelvic bone marrow: serial MR imaging before, during, and after radiation therapy. Radiology 1995;194:537–43.

[16] Stevens SK, Moore SG, Kaplan ID. Early and late bone-marrow changes after irradiation: MR evaluation. AJR Am J Roentgenol 1990;154:745–50.

[17] Kauczor HU, Dietl B, Brix G, et al. Fatty replacement of bone marrow after radiation therapy for Hodgkin disease: quantification with chemical shift imaging. J Magn Reson Imaging 1993;3:575–80.

[18] Mumber MP, Greven KM, Haygood TM. Pelvic insufficiency fractures associated with radiation atrophy: clinical recognition and diagnostic evaluation. Skeletal Radiol 1997;26:94–9.

[19] Fu AL, Greven KM, Maruyama Y. Radiation osteitis and insufficiency fractures after pelvic irradiation for gynecologic malignancies. Am J Clin Oncol 1994;17:248–54.

[20] Weber M, Hasler P, Gerber H. Insufficiency fractures of the sacrum. Twenty cases and review of the literature. Spine 1993;18:2507–12.

[21] Richardson ML, Zink-Brody GC, Patten RM, et al. MR characterization of post-irradiation soft tissue edema. Skeletal Radiol 1996;25:537–43.

[22] Biondetti PR, Ehman RL. Soft-tissue sarcomas: use of textural patterns in skeletal muscle as a diagnostic feature in postoperative MR imaging. Radiology 1992;183:845–8.

[23] Paz IB, Wagman LD, Terz JJ, et al. Extended indications for functional limb-sparing surgery in extremity sarcoma using complex reconstruction. Arch Surg 1992;127:1278–81.

[24] Fox MG, Bancroft LW, Peterson JJ, et al. MR imaging appearance of myocutaneous flaps commonly used in orthopedic reconstructive surgery. AJR Am J Roentgenol, in press.

RADIOLOGIC
CLINICS
OF NORTH AMERICA

Radiol Clin N Am 44 (2006) 473–478

Index

Note: Page numbers of article titles are in **boldface** type.

doi:10.1016/S0033-8389(06)00042-X

Changing Your Address?

Make sure your subscription changes too! When you notify us of your new address, you can help make our job easier by including an exact copy of your Clinics label number with your old address (see illustration below.) This number identifies you to our computer system and will speed the processing of your address change. Please be sure this label number accompanies your old address and your corrected address—you can send an old Clinics label with your number on it or just copy it exactly and send it to the address listed below.

We appreciate your help in our attempt to give you continuous coverage. Thank you.

W. B. Saunders Company

SHIPPING AND RECEIVING DEPTS.
151 BENIGNO BLVD.
BELLMAWR, N.J. 08031

SECOND CLASS POSTAGE
PAID AT BELLMAWR, N.J.

This is your copy of the
_____ CLINICS OF NORTH AMERICA

00503570 DOE—J32400   101   NH   8102

JOHN C DOE MD
324 SAMSON ST
BERLIN   NH   03570

XP-D11494

JAN ISSUE

Your Clinics Label Number

Copy it exactly or send your label along with your address to:
W.B. Saunders Company, Customer Service
Orlando, FL 32887-4800
Call Toll Free 1-800-654-2452

Please allow four to six weeks for delivery of new subscriptions and for processing address changes.